Modeling The
UNIVERSE

By
R.A. Kirby PhD

 FriesenPress

One Printers Way
Altona, MB R0G 0B0
Canada

www.friesenpress.com

ISBN
978-1-03-917823-6 (Hardcover)
978-1-03-917822-9 (Paperback)
978-1-03-917824-3 (eBook)

1. SCIENCE, SPACE SCIENCE, SPACE EXPLORATION

Distributed to the trade by The Ingram Book Company

Table of Contents

This book is dedicated to my family past present and future.

PROLOGUE

This is a modern Voyage of the Beagle that takes the reader upon a journey of discovery into the mechanics of our universe. The Beagle took Charles Darwin into the rivers and inlets around Terra del Fuego and changed science. The current voyage describes our universe in a new and direct manner that disposes of old ideas. Our voyage answers the most basic questions about our universe. The previously convoluted story becomes a simple familiar tale of nature. The modeler's visual perspective is used to create simple direct images that explain known facts. Our voyage moves the reader from the smallest bits of matter to the largest. Fundamental particles are modeled, black holes are broken and Quantum Mechanics is revealed. Extra dimensions, spooky action at a distance and curved space are explained. This image based description of our universe gives the reader direct explanations of simple facts.

INTRODUCTION

Modern physics is in need of a re-boot resulting in a new description of reality. According to Lee Smolin as outlined in his book, The Trouble with Physics (references), modern science is in need of new perspective and a new solution to its current mess. This he asserts is due to a loss of perspective that has accompanied String Theory. Current physics has mathematical description of some observations but lacks a credible theoretical explanation for experimental results.

This book describes a *Theory of Matter* that explains the physical observations obtained by modern physics. The result is a new version of the story of matter and how it makes our universe. Our species would like to know the answer to the question, what is the universe made from? One answer comes from Lawrence W. Krauss in his book A Universe from Nothing (references). This is close to the answer that is exposed in the chapters that follow. The current tale is in the vein of the oldest stories. It describes a journey to explore new lands followed by a return home. Genuine accounts of travels like those of Marco Polo and Ibn Batuta have influenced generations. Versions of travel tales were told around campfires and these have seeded our literature. Stories of a journey and the return home occupy our literature and include the travels of Odysseus, Tom Sawyer and Bilbo Baggins. The current story is a journey of discovery into the heart of matter and to the birth and death of our universe. The transport for this journey is a scientific model of the fundamental particle of matter. The result is a valid *Theory of Matter* that is a survival tool for a technological species. The model provides a theoretical underpinning for the computational and experimental results in the Standard Model. Knowledge of the nature of matter leads to a new interpretation of cosmology and brings us closer to home.

Charles Darwin said in his introduction to The Origin of Species (references) that

> *In scientific investigations it is permitted to invent any hypothesis and*
> *if it explains large and independent classes of facts, it rises to the rank*
> *of a well rounded theory.*

The voyage of discovery in this book is revealed through the modeling pathways that produced the description of matter. Once revealed the model *explains large and independent classes of facts.*

Like the expeditions made by Ibn Batuta across Northern Africa, preparation in advance of departure is essential. Our journey to the heart of matter requires the production and refinement of a model. A first step is to briefly introduce the subject by describing the physics that is background. The modeling exercise involves producing and testing the model against known observations. The model is then used to explain and study the universe.

The properties of matter are required input into the model and their modeling takes up the first four chapters. The three chapters that follow test and use the model to describe Quantum Mechanics, Relativity and Time. The next two chapters deal with the cosmology modeling that emerges from the particle model. A final chapter looks at the universe in which we live.

A starting place for this adventure is where modern physics ends. This means that a short summary of that subject is essential. The information here is widely available on many sources and is referred to in the following chapters. The Standard Model holds the accumulated knowledge of modern physics. A part of this wisdom is held in a table of fundamental particles that has been uncovered through collision experiments. The matter particles listed in the Standard Model table are concluded to be fundamental. This is possible because these particles decay into lower energy products through a variety of pathways.

The Standard Model table of particles includes the quarks, leptons and neutrinos that make up matter. The quarks and leptons come in three generations and carry charge while the neutrinos are neutral. The table also has the bosons that mediate interactions between the fermions. These include photons, gluons and the weak force boson intermediaries. The gluons, photons and neutrinos are neutral and travel at the speed of light. The odd seventeenth entry in the table represents the Higgs boson that has been hypothesized to be responsible for inertial mass. The gravitational interaction is explained by the Standard Model to be due to gravitons that have never been observed.

The Standard Model also has an equation that calculates a LaGrangian Energy for matter particles. The term LaGrangian refers to a calculations for interactions assuming particles are frozen in time and place. The full equation encapsulates the accumulated experimental knowledge of particle interactions. The empirical nature of the equation is exemplified by Maxwell's calculation of the speed of light. This is given by the relationship $c_o = 1/(\mu_o, \varepsilon_0)^{0.5}$ where μ_o and ε_0 are experimental constants. Einstein's Field Equations are also included and are based upon Newton and Galileo.

The Standard Model does not produce an underlying theoretical explanation for observations and calculations. That deficiency has lead to the proliferation of theories that cannot be included into the standard Model due to their speculative nature. Quantum Field Theory (QFT) concludes that matter is an excitation in a field that exists in all of space. This is elusive by being vague about the source of each field that is linked to each particle. This concept is clear inspiration for the Higgs Hypothesis that proposes an explanation for inertial mass. That guess explains inertial mass for matter particles as an interaction with a Higgs Field. This unusual field purports to permeate all of empty space and to have a non-zero energy value everywhere. The Higgs hypothesis does not explain the source of this field or where the energy for this field comes from. The 2012 Nobel Prize in Physics was given for the discovery of the Higgs particle. This highly unstable particle has a mass of 125 GeV and a lifetime of 10^{-22} seconds. The existence of a highly unstable matter particle does not give evidence of a field interacting with all matter particles.

String Theory remains outside the Standard Model due to its speculative nature. This guess upon the nature of matter calls itself a theory. A mathematical hypothesis that holds matter to be due to vibrations in empty space, the 'Theory' admits that it makes no predictions available to experiment. The idea is that matter particles are strings of energy existing at the Planck scale making it unfalsafiable. The basic belief, without evidence, is that empty space is mutable and can store energy in its distortions. String Theory is a form of mathematical mysticism never inspired by experiment and originally thought to give a description of the Strong Interaction. The Strong Interaction was better represented by Quantum Chromo Dynamics and

String Theory was converted to its present usage as a theory of matter. This idea falls on the rocks by lacking any experimental evidence to support the concept.

The following description of nature provides a more plausible explanation for our universe. Some of the proposed descriptions of matter can have value and fragments of truth can be found in each. QFT is right and matter particles are quantum fields but more than this general description is needed. String Theory is correct about strings of energy, but wrong in the idea that empty space is mutable and can store energy. The failure of modern physics is written large in the list of questions that remain unanswered. These include the Double Slit experiment, Quantum Tunneling, Quantum Entanglement and Wave Function Collapse. The entire field of Cosmology depends upon particle physics and cannot explain dark matter, dark energy and the expansion of our universe.

Chapter 1

CREATING A MODEL

Richard Feynman's view upon creating a model of the fundamental particle was clear when he said, (references).

> the more you see how nature behaves, the harder it is to make a model that explains how the simplest phenomena actually work,. So, theoretical physics has given up on that.

This revolution starts here because the vehicle for this journey is a model of the fundamental particle of matter. Preparation for the journey requires construction of the model just as Homer tells us that "black ships to Troy they steer" ..."supplied by Agamemnon's care". Ibn Batuta, the Muslim scalar who crossed North Africa in the 14th Century, knew the routine of preparing for travel. A physical model of matter is our "ship of the dessert" and will be our key to the universe.

A three dimensional model must explain the emergent properties of mass, charge and spin. It should provide physical mechanisms for particle upon particle interactions. The model should be expressed in physical and mathematical form. Knowledge of the construction material making up matter is therefore crucial.

This means that the first step in modeling matter and the universe is to know what matter is made from. There is in fact a body of evidence supporting the *"light is fundamental"* hypothesis. George FitzGerald, the Irish physicist, as early as 1880, suggested that electromagnetic ether permeated all of space. He thought that matter was a knot of electromagnetic energy. There was insufficient evidence and even when evidence arrived the idea was never widely accepted. There is circumstantial evidence of this as particles weight more after they have absorbed light and weight less after emission. Further it can be shown that particles and anti-particles annihilate one another upon contact to release all of their energy as light. Particles also emerge from a volume of space that is subject to a strong electric field. Virtual particles from the background exist for short periods of time and are seen in the Lamb Shift. This experimental result is explained as hyperfine splitting of transition energies in atoms due to the presence of virtual particles.

There are two experimental facts that make it impossible to dispute that matter is made from light. The first of these is in the Lorentz Transformations and the second is the "Double Slit" experiment. These two results are fully verified experimentally. The Double Slit will be described in the chapter on Quantum Mechanics. The Lorentz Transformations are a set of equations that convert from one inertial frame to another. They were put forward by Hedrik Lorentz as an explanation for the failure of the Michaelson and Morely experiment (reference).

These transformations relate space and time coordinates for two systems that are moving at a constant velocity relative to one another. One of these velocities is the speed of light and the second is the velocity of the moving object. The finite speed of light is an important part of this and was determined by the Danish Astronomer Ole Roemer in1676. This was a revelation to the common view that the speed of light was infinite as assumed by Galileo and Newton. The Lorentz and the Galilean transformations are shown in Table 1.1 below.

GALILEAN	LORENTZ
t/t=1	$t'/t = \dfrac{1}{\sqrt{1 - v^2/c^2}}$
X/X'=x+vt	$x'/x = \dfrac{1}{\sqrt{1 - v^2/c^2}}$
y'=y'	y'=y'
z=z'	z=z'

TABLE 1.1 Galilean and Lorentz Transformations where v is velocity, t, is time and C is the speed of light.

The origin of the Lorentz Transformations is important information because it clarifies how they work. The equations come from a practice problem given to students by Hedrik Lorenz (1853-1928 Dutch Physics Professor). The study problem could have been given by Euclid or Pythagoras in Ancient Greece. The problem is a river crossing as in Figure 1.1 below where a swimmer must cross a flowing stream. The swimmer must cross the flowing river and then swim a distance up the stream. This is followed by a return trip back to the near shore and a swim back to the starting place. The question for the student required the use of the Pythagorean Theorem to determine the time and distance travelled. The velocity of the swimmer and the stream are superposed upon one another. The resulting distance and time both depend upon the two velocities. The velocity of the river determines the lengthening of the distance and time for a circuit by the swimmer. In the Lorentz Transformations the flowing water takes the place of a moving object and the speed of the swimmer replaces light moving at C. This is depicted in Figure 1.1.

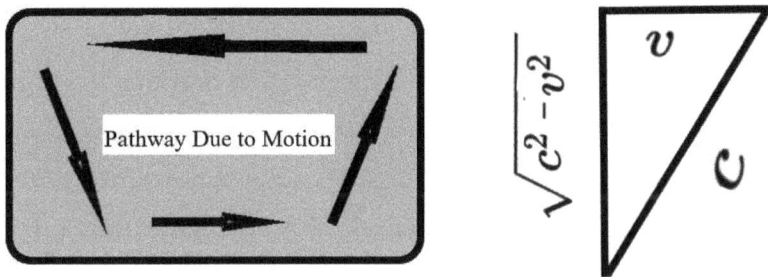

FIGURE 1.1: River Crossing and Light Path Model

The journey around a circuit is dictated by the constant speed of the flowing river that is imposed upon the constant velocity of the swimmer. The parallel between the swimmer and the motion of light is the root of the Lorentz Transformations. The solution of this problem yields an important relation that permeates physics. This factor is called **y** and gives the transformation between the two inertial frames of reference. The value of **y** tells us how distance and therefore time is affected by the superposition of these two constants.

$$\gamma = \frac{1}{\sqrt{1 - v^2/c^2}}$$

Therefore, when the velocity of the river or motion of an object equals zero the Lorentz Transformations collapse to those of Galileo in Table 1.1.

The extended circuit experienced by light inside a moving particle also changes the mass, energy and apparent length of moving objects. These changes also are described by the **y** value as in Table 1.2.

GALILEAN	RELATIVISTIC
E=MC²	$E = \dfrac{MC^2}{\sqrt{1 - v^2/c^2}}$
M/ M'=1	$M/M' = \dfrac{1}{\sqrt{1 - v^2/c^2}}$
L/ L'=1	$L'/L = \sqrt{1 - v^2/c^2}$

TABLE 1.2 Galilean and Relativistic Energy, Mass and Length.

The mass of a moving object changes as a result of motion and this alters the energy of the particle. The energy required to move an object is known to all and will be further elucidated in the chapter that follows. The apparent length contraction due to motion of a moving object is a light effect. Here the light coming from the front and back of a moving object arrives nearly simultaneously to make the appearance of length contraction.

Lorentz proposed these equations as a solution to the apparent failure of the Michaelson-Morely experiment. They did not save the theory, but were used by Einstein in his 1920 publication (references). There he stated that these equations were used because they *gave the correct answer.*

The reason Einstein was eager to accept the Lorentz relations came from experimental verification. The mass increase due to motion was demonstrated as early as 1904 and is seen in Figure 1.2.

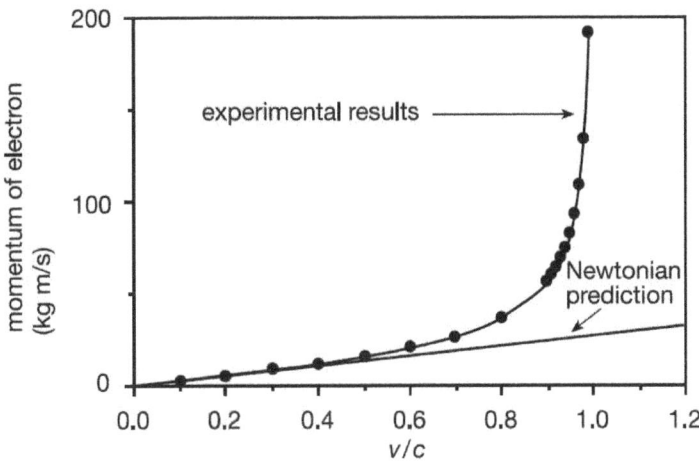

FIGURE 1.2. Newtonian verse relativistic increase in mass of electrons as a function of velocity, v where p is momentum, m is mass and c the speed of light. Walter Kaufman (1901-1903) (reference).

Experimental results like this made it a black letter fact that Lorentz had the correct answer. This result for an electron accelerated in an electric field was available to Einstein. Time dilation has also been verified in many examples with one of these evident in the lifetime of fast moving muons. The conclusion from this is that matter is made from light. That light circles within the body of a matter particle to give us the **y** relation. The Double Slit Experiment also affirms that matter is made of light. The hypothesis is that,

Light is the fundamental building block from which all matter is constructed.

Armed with the fact that matter is a form of electromagnetic radiation, it remains to describe that light. In advance of a description of the light that must occupy matter, a description of the background environment is useful. The situation is that a background electromagnetic environment exists in empty space. Light is an electromagnetic harmonic oscillator that transfers energy by oscillating between magnetic and electric fields. This oscillation, like a simple pendulum clock, is never fully silent. Even in the coldest space very low levels of light flicker and move across the void. The temperature of empty space is a measure of its radiation content and has been determined to be below 3.7 ° K in the hot empty space in our universe. Just outside the Swarszschild Radius of a black hole the temperatures are estimated to be below 10^{-9} Kelvin but never zero. The background field is always a source or a sink of energy and a highway for energy transfer.

The result is that matter particles are made from light and are imbedded in an electromagnetic background. Light, moving at speed C within the body of matter, can achieve this feat if it moves in a closed path. This can only be true if high frequency radiation has its electric and magnetic fields pushed together to produce a closed loop. The overlay of fields produces attraction that can pull the light path into a circular motion. The wavelength of such light must be very short to fit within particles that have diameters below 10^{-18} M. The electromotive force that changes light into matter is due to overlap of virtual photon flow that stabilizes circular light pathways.

The properties associated with fermions like mass, charge and spin emerge when light circles in a path at C inside the matter particle. In effect high frequency light condenses into matter as an emergent property that arises from self attraction. Different particles incorporate different combinations of high frequency light. This produces separate properties for matter particles creating the attributes of mass, charge and spin. Each particle represents an energy minimum in a three dimension surface.

Stimulated emission lasers tell us that light is self attraction. This means that attractive potential correlates with energy density. The result is that light at high frequencies always has a light pathway that moves in closed loops. An artistic image of the light that moves inside matter is in Figure 1.3 below.

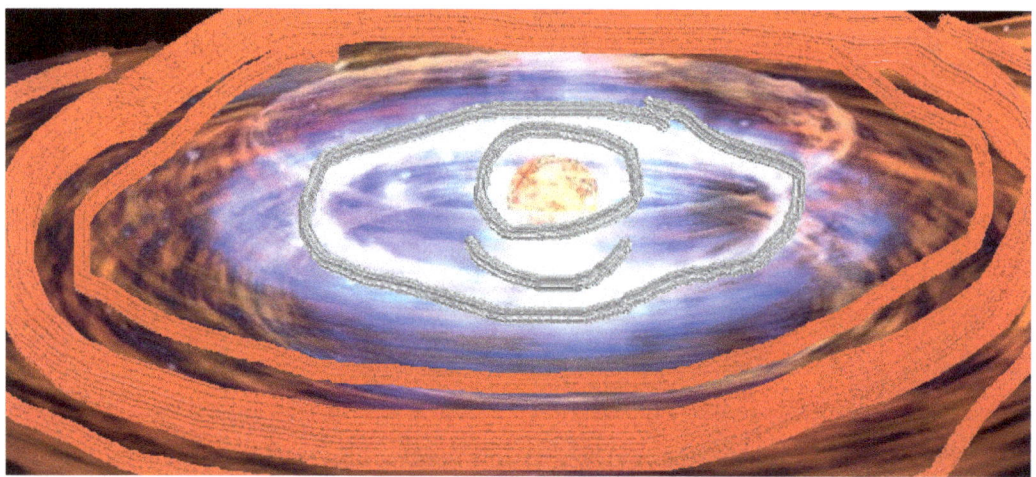

FIGURE 1.3 Image of the Light Inside a Fundamental Particle.

This depiction places the hottest radiation towards the center of the moiety and larger red circles away from the center. The sticky nature of high frequency light means that all the light collected within a volume of space is self attractive.

The circling light inside the particle makes a path that takes as dew as ~10^{-27} seconds. The light spreads to fill that volume allowed by any containment fields. A field of light spreads to occupy the available space just as a liquid like water that is self attractive and yet spreads to fill a volume. A very high resolution image of a matter particle would show a spread of contributions. The loss of resolution resolves a particle into a sphere of light. Figure 1.4.

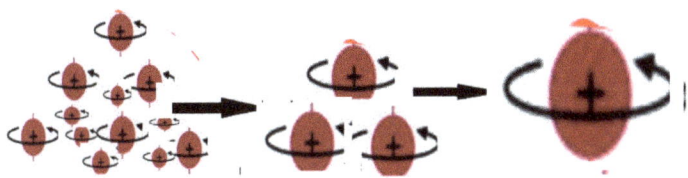

FIGURE 1.4 High, Medium and Low Resolution of the Light in a Particle Field.

The light travels in pathways that have an even number of wavelengths in a circuit to avoid destructive interference. The light inside of matter has a Fourier Transform of its radiation content showing separate contributions as in Figure 1.5. Each particle has its own combination of light and therefore a fingerprint Fourier transform.

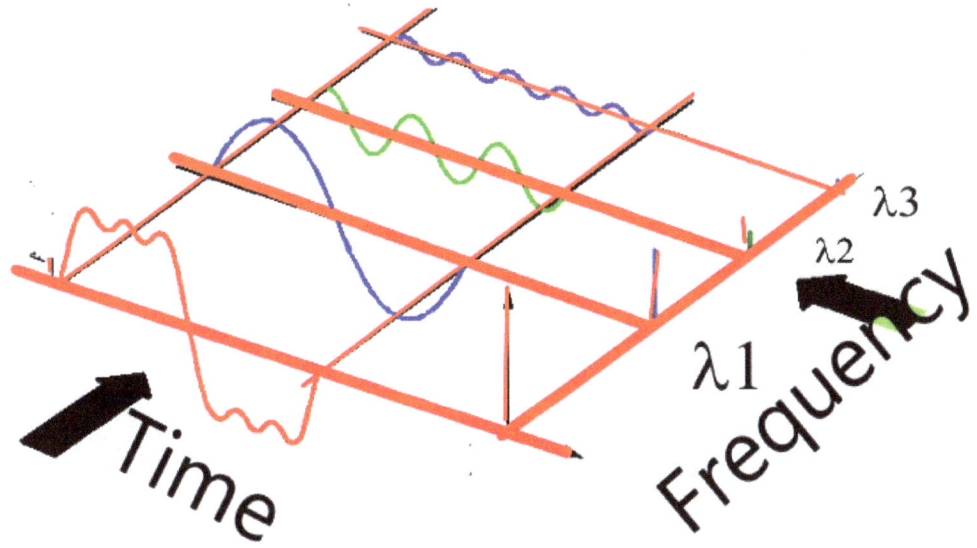

FIGURE 1.5. Hypothetical Fourier Decomposition of Light in Matter

The hypothetical time and frequency domain plots in Figure 1.5 show a founding high frequency burst surrounded by longer wavelength contributions. Massive particles are unstable in the absence of high temperatures and pressures and decompose rapidly. This occurs through a separation of the light components of a matter particle to produce the new entities.

A particle is therefore a pool of light flowing within a confined volume. Hot light with the highest frequency is most sticky and forms particles. The cold light, like that from our sun, flows through open space at speed C without forming into matter. These two forms of light are cartooned in Figure 1.6.

 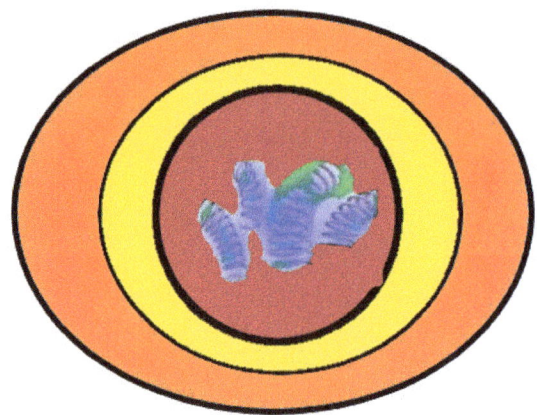

FIGURE 1.6 Light (photons left and particle right).

The difference between the light in a particle and that in a photon is in the frequency of the light making each. The light in matter can be made to move by the addition of radiation into the body of matter that translates the entire volume.

This means that a mathematical representation of matter requires a representation of its light content. There are a number of mathematical functions that can do the job of representing light for the model. A computational representation of the light that occupies fermions would allow the mapping of electromagnetic fields that overlap. Studies searching the frequency range for the matter particle have the potential for determining the resonant frequencies for individual particles. The relationship between the frequency of light and its self attractive potential requires modeling. Experimental studies with high energy radiation can supplement the mathematical representation. A sine function is the basic form of the equation for the propagation of light in space. A mathematical function is needed that correlates the light upon light interaction with frequency. Overlapping fields and particle decay sequences can teach the mathematical model to improve the representation of matter particles. A proposed mathematical form to represent the light making up a particle is in Equation 1.1.

Equation 1.1 $Mass/C^2 = (U((Sin\vartheta + Cos\Omega)_{XYZ})_{1\text{-}5})^2$

Here the $(Sin\vartheta + Cos\Omega)$ term gives the shape of the light with three X, Y and Z contributions and 1 through 5 harmonics of the $(Sin\vartheta + Cos\Omega)$ radiation. The function representing light upon light attraction, U, is based upon experiment.

The computational version of a particle becomes the overlay of several individual contributions due to different harmonic frequencies. A computer search would be necessary where electromagnetic contributions to the overall radiant energy are varied and stability is measured by lowering the energy content. A model of a particle is built up by an overlay of contributions having different wavelengths

The idea that light makes up matter is foreign since we experience light and matter our differently. However, the evidence is absolute and matter is a form of light. The properties of matter can therefore be explained within the description of its light content. This is the direction of the journey before us and *there are miles to go*

Chapter 2
HOW MATTER MOVES

Motion is the most basic property of matter and has been guessed at since the earliest time. The Greeks thought that matter had agency and moved according to an internal desire. This echoes of animism as a view of nature. Newtonian Mechanics told us that acceleration and mass were mathematically related. The lesson was that external forces were responsible for motion. Ernst Mack (1838-1926) suggested that the universe interacted through gravity to give matter its inertial mass. Newton thought that matter was, as Democratus had suggested, small immutable balls of solid substance. Special Relativity used the Lorentz Transformations as described above but avoided any description of matter. The Schrödinger Equation treats matter as a point charge with mass and this is the default position of the Standard Model. Modern Physics in the form of QED treats matter a point charge that moves by simple translation. A model must therefore explain how particles made from light move through space.

The answer to the question is that matter moves as the light inside the matter moves. The light path in matter is stretched by motion as described above. When matter is stationary then its light can take a circular path and this is stretched upon motion. A macro world analogue for the motion of the light inside of matter is in a top that spins on an axis while moving across the floor.

Light interacts with itself through addition and can have positive or negative interference as in Figure 2.1.

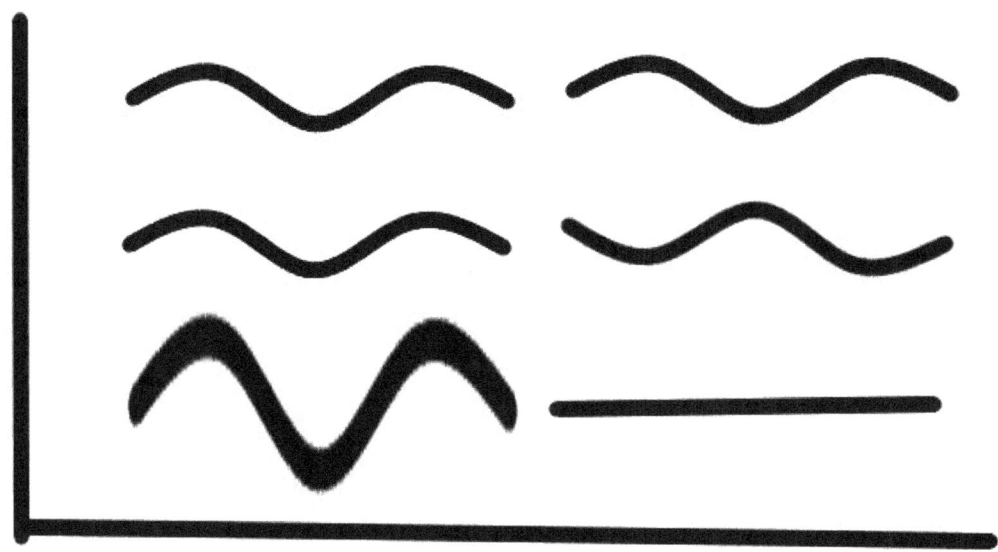

FIGURE 2.1 Light Interference: Amplitude Modulation

Different frequencies also are additive to produce complex wave shapes. The result is that, if matter is a bundle of light, then to make it move interference of some form is necessary. This means that pushing an object involves transferring light into the object. This is consistent with the fact that energy is required to produce motion. This is the source of inertial mass that resists motion even in the absence of any opposing force. Acceleration requires the introduction of light into the matter being moved.

A particle moving in an inertial frame changes velocity when light is added to or subtracted from the moving particle. Energy moves through space as radiation when particles exchange light. When fields overlap then virtual photons flow and move energy from hot to cold under the Second Law of Thermodynamics. Any addition of energy results in new motion that superimposes itself onto all other motion. The motion is recorded in the stretching of the light path within the particle that is superposed upon other stretching. The trace of the light inside of matter changes as energy is absorbed and emitted. The light trace changes are depicted for different modes of motion as represented in Figure 2.2.

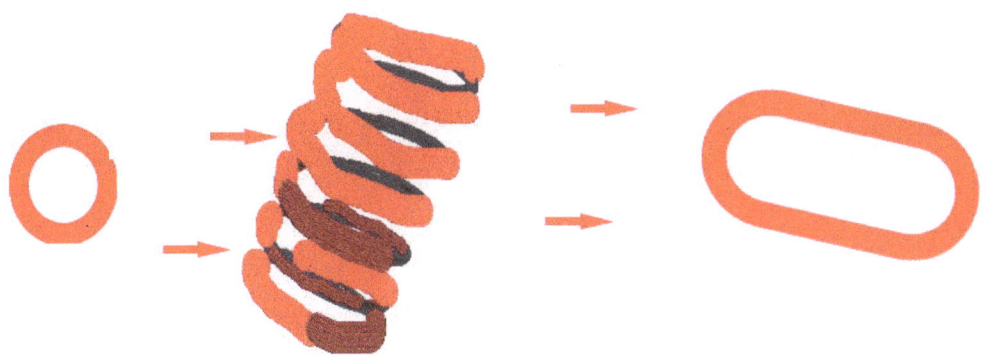

FIGURE 2.2 Light Trace Due to Motion.

The light can circle in closed or open loops as the matter particle moves. Each additional motion is superimposed upon the total velocity and direction of the light moving inside the matter.

Whether one is heating an object or pulling it with a horse, then all forms of work involve putting energy into the particle. Light enters the energy matrix of a matter particle and produces motion and direction. Putting energy into matter produces a Relativistic Mass increase as light. The energy stored in the matter particle upon motion is the Kinetic Energy. This energy has been experimentally determined as $Ke=1/2\ MV^2$ where M is the mass and V the velocity.

The effect of the addition of energy to the light making up matter is modeled in Figure 2.4 below. The extension of the light path is a function of both the speed of light and the velocity of the matter. This means that the light inside of matter is in a fundamentally different energy state when in motion than when stationary.

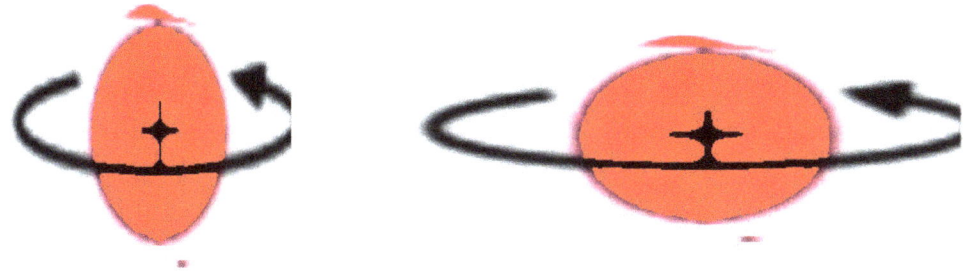

FIGURE 2.3 Elongation of light pathway along the axis of Motion

A macro world analogue of the light inside the fundamental particle is modeled by an elastic band. Stretching of the elastic requires energy and the stored energy is recovered when the tension is released. Moving any object or changing a direction requires the constructive addition or destructive subtraction of energy. Consider a room that is bathed in white light and green light is added. The room can be made to appear green by the addition of light having a wavelength of 550 nm. Increasing the amount of green light as a percentage of total light will cause the room to look green. The same principle applies when adding energy to the light inside of matter and making it move. When a force moves one kilogram then it takes double the force to move two kilograms. The resistance to motion scales with the amount of matter and therefore the total amount of light energy.

Einstein said in 1907 that his happiest day was the one in which he realized that inertial mass and gravitational mass were equivalent. He then went on to use the Equivalency Principle to make a series of predictions about matter and light in gravitational fields. Actually, the equivalence between these two masses had been known in Newton's time and had been confirmed at a high level of precision by baron Roland von Eotvos around 1885. The baron used a tortional measure to check with high accuracy the equivalence principal. Eotvos was cited by Einstein in his 1916 paper. Einstein used this principal to predict light bending in a gravitational field. The position taken here is that the Equivalence Principle applies for all forms of acceleration. The result is that all acceleration fields have the same effect upon the light pathway inside the matter particle. This means that placing matter in a strong electric field has the same effect as motional acceleration. The gravity field of earth is now stretching the radiation pathways inside our quarks and leptons. This action stores energy as gravitational potential energy given by the equation $E_{pot} =$ mgh. The stored energy, as described by this small equation, is due to the mass (m), the strength of the gravity field (g) and the height above the center of mass creating that field (h). Gravitational energy can be converted into Kinetic Energy by release of the matter to fall towards the center of the gravity well. The general situation for the internal geometry of matter experiencing any form of acceleration is depicted in Figure 2.4 below. Here the Y axis represents the strength of the acceleration field and the X axis as matter is moved into and out of that field.

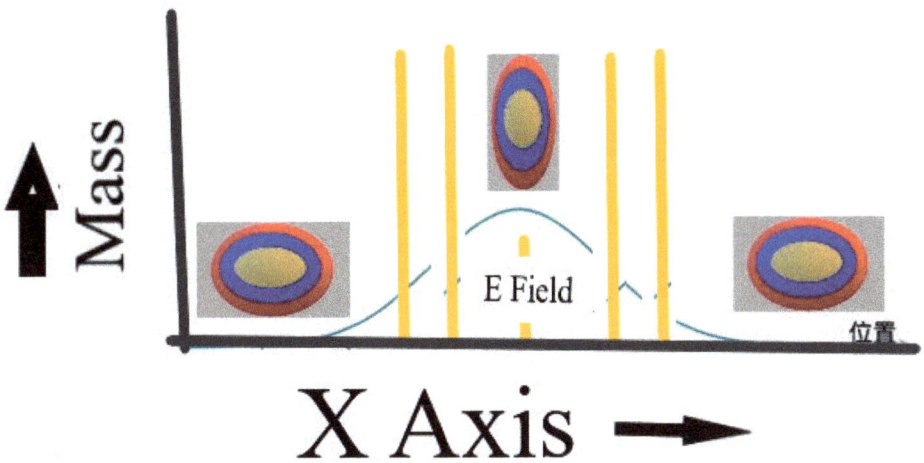

FIGURE 2.4. Light Path Shape as a Function of General Acceleration.

The interaction between the light in matter and the light in photons can change the trajectory of moving particles. The situation where the wavelength of light is not suitable for interaction then interaction does not occur. The case with light of appropriate wavelength means the light in the moving matter interacts with the light in the photons. The scattered matter due to interaction with light has a wavelength that is changed by the collision. This is the Compton experiment where photons have a directional energy that is transferred to the electrons. The result is that the electrons have their trajectories changed.

A particle held in an acceleration field has its radiation pathway extended. This is achieved by adding energy to the light matrix of the entity. This knowledge can be used to help us understand the gravitational field. Acceleration inserts energy into the body of the particle and deceleration extracts energy from that particle. Therefore, placing matter into a gravity field distorts the radiation pathway and stores energy as in Figure 2.4 above. When an apple falls from a tree then stored gravitational potential energy is converted into kinetic energy. A charged particle held in an electromagnetic field and then released converts stored potential energy into kinetic energy. Table 2.1 documents the parallel situations for acceleration fields.

FIELD	INERTIAL ACCELERATION	GRAVITATIONAL ACCELERATION	ELECTROMAGNETIC ACCELERATION	GLUON FIELD ACCELERATION
Stationary State	No Motion Radiation Pathway is Minimal	No gravity field, Pathway is Minimal	No Electric or magnetic fields. Pathway is Minimal	Subatomic matter always moving and in a field. Pathway is Variable
Accelerating All Changes in Motion including rotational motion.	Matter is being accelerated Radiation Pathway is Stretching	Matter is entering a gravity field Pathway is Stretching	Charged Particle being acceler-ated. Pathway is Stretching	Charged Particle being acceler-ated. Pathway is Stretching
Constant Field	Matter moves at constant velocity Stretched Pathway remains constant	Matter is held in place Stretched Pathway remains constant	Charged Particles held in place in an EM field. Stretched Pathway remains constant	Charged Particles held in place in a Gluon Field Stretched Pathway remains constant

TABLE 2.1 Acceleration Field Equivalencies.

A gravity or electrical field superposes its influence upon any other motion to produce a net motion.

Matter therefore moves as its light reorganizes along the direction of motion as the light inside of matter reorganizes along a trajectory. The light inside the matter particle always moves faster than macro translation. The addition of light having directional component delivers motion to the bundle of light making up matter. The net effect is that all acceleration fields add energy to matter particles.

This explanation for the motion of matter is in conflict with that of the Standard Model. This interpretation is that that matter translates like a macro world entity that simply moves as a solid unit through empty space. The Higgs explanation for inertial mass assumes that an immutable particle moves through space like a beach ball through water. Particles then interact with a hypothetical field like a ball interacts with water or air molecules. The next chapter describes the fundamental properties of matter that we know as charge mass and spin.

Chapter 3
INTRINSIC PROPERTIES MASS, SPIN CHARGE

The Higgs mechanism for inertial mass is the Standard Model explanation for this property. The other two properties of matter exist in the absence of any mechanistic explanation. The fact that the Standard Model treats matter particles as black boxes has given us a wrong answer coupled with two non answers. The current chapter corrects this deficiency by delivering mechanisms for these three emergent properties of matter. The simple answer for all three mechanisms comes from the nature of the light that makes up matter.

MASS

A body placed in empty space requires energy as input before motion can occur. This is true even in the absence of any friction or field that could impede motion. This is the inertial mass of matter and is designated as the property that makes matter resists changes to motion. The previous chapter has shown that matter moves by external influences that transfer energy to the light making up the matter particle.

The Standard Model lists the Higgs boson in the Table of Fundamental Particles. This implies acceptance of the Higgs Hypothesis that inertial mass is due to interaction with a field that permeates all of space. The idea treats matter particles as black boxes and uses the Quantum Field hypothesis to propose an interaction between particles and a mysterious Higgs field. This idea fails as easily as the question regarding the Equivalence Principle and gravitational mass. If inertial mass is due to interaction with a Higgs Field upon motion, then why does a gravity field generate mass increase when matter is stationary? The Higgs Hypothesis fails Accum's test by the complexity of having each particle type interact differently with a field. The problem with this is in the idea that a field having non-zero energy exists in all places in the universe. This complex explanation for inertial mass does not explain how matter interacts with the Higgs field.

Inertial Mass is the resistance to all forms of acceleration as shown in the previous chapter. Try to lift an object in a gravity field or to move an object in the absence of resistance exposes this fundamental property of matter. The current model of matter particles indicates that resistance to motion comes from the requirement of energy transfer into the electromagnetic matrix of the particle. All forms of acceleration add energy into the particle body and change the light pathway. The addition of energy to the light inside of matter expresses itself as relativistic mass increase. The Lorentz physics problem described above demonstrates that a lengthened path correlates with time dilation and mass increase. The amount of light energy in a particle scales with the amount of resistance that particle has to acceleration. The result is that it takes more energy to move a larger mass.

The equation for the energy of a photon of light is, $E=h\lambda$, and the energy of a particle is given by the familiar equation $E=MC^2$. In these equations E is the energy, M is the mass, h is Planck's constant and λ is the wavelength. Combining these two, now that we know they are made from the same substance, gives $M= h\lambda/C^2$. This relation was first presented by Louis DeBrogalie (1892-1987). The result means that mass, scales with the frequency of light with all other terms as constants. The resistance to motion is a scalar function of the number of wavelengths per second and therefore the energy. The result is that high frequency radiation requires more energy to move than lower frequency light. Objects move when energy is inserted into the light matrix of the fundamental particles that make up matter. Therefore, all motion must overcome the resistance to motion dictated by the amount of light that is being accelerated. The simple result that known to all is that mall masses move more easily than large amounts of matter.

Mass is therefore a scalar measurement of the total amount of electromagnetic radiation that is contained within the particle. The different fermions in the Standard model table of fundamental particles each contain an individual mix of radiate energy. Each aliquot of matter is nature's solution to an energy calculation that finds energy minima in the mix of contributing frequencies.

The stable matter particles in our universe come from the decay of high energy unstable particles. Matter therefore collects in pools of stability after decay into lower energy minima as energy descends a ladder of possible states. The parallel between the decay of high energy matter into lower energy analogues is parallel to a water fall. Here the water collects in pools that then feed into other lower energy states as depicted in Figure 3.1.

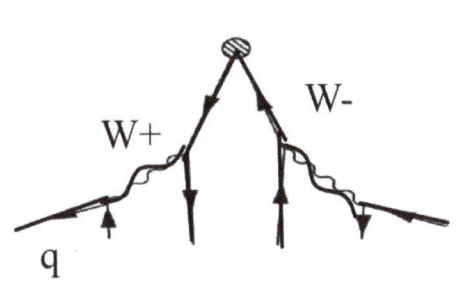

FIGURE 3.1 Mass Particles and Pools of Water.

Collision experiments smash protons together releasing the kinetic energy in quarks and this energy is available to form new matter particles. The high frequency light released in these collisions is available to make new particles. The quarks and leptons that make the matter that we experience are listed in Table 3.1 and 3.2 below.

QUARK FLAVOR	MASS (GEV)/C^2	ELECRIC CHARGE
Up	0.005	+2/3
Down	0.01	-1/2
Charm	1.5	+2/3
Strange	0.2	-1/2
Top	180	+2/3
Bottom	4.7	-1/2

TABLE 3.1 Quark Masses and Charges.

LEPTON	MASS (GEV)/C²	CHARGE
Electron	0.511	-1
Muon	106	-1
Tauon	1,780	-1

TABLE 3.2 Lepton Masses and Charges.

The three layers of matter in the quark and lepton tables reveal the nature of matter. The three orders of matter are almost certainly related to symmetry around the X,Y and Z axes. In a three dimensional universe the efficient use of space becomes a driving factor in matter stability. The packing of light pathways in three dimensional spaces dictates the energies and shapes of the three orders of matter. The lowest energy path will be symmetric about three axes; the second generation has greater energy with symmetry around two axes. Finally, the most energetic and unstable analogues have symmetry around a single axis. The higher mass entries have higher frequency radiation and a greater amount of total radiant energy. Decay of the higher order particles to their lower energy analogues happens as their composite light explores and finds lower energy states. The high mass particles decay through multiple routes to lower energy analogues.

Computational studies can be used to discover the wavelengths of light making up a series of matter particles. This can be done for a group such as the leptons in Table 3.2. This task is outside the reach of the current method of representing matter particles found in the Schrödinger Equation.

CHARGE

The situation for charges and electromagnetic interactions differs from that of inertial mass because there are no theoretical explanations here. The Standard Model does not have a mechanism explaining what causes matter to have charge. The route through this area of physics begins with a basic description of the experimental results. This will then be followed by a first ever explanation for charges in matter and the electromagnetic fields that emerge.

The physical results of charged matter were described by Michael Faraday (1791-1867) and James Clerk Maxwell (1837-1879). Charges and their fields have never had their underlying physics described. The experimental side of electromagnetism can be summarized by a few commonly available images. A summary of charge interactions are shown in Figure 3.2 below where electric and magnetic fluxes are depicted.

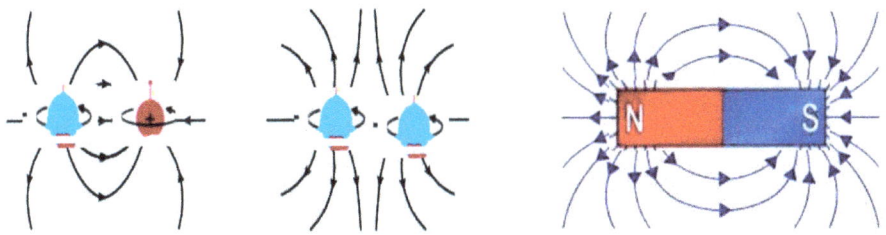

FIGURE 3.2 Positive and Negative Charges and a Bar Magnetic.

The fundamental fact is that charged particles are either repulsive or attractive as given in Figure 3.2. The force lines emanating from charges and around a bar magnet are in Figure 3.2. The charge upon charge interactions represented in Figure 3.2 have a force given by the equation $F=Kq_1q_2/r_{12}^2$. Here q_1 and q_2 are the magnitude of charge, r is their separation and K is the coulomb constant. The force between two current carrying wires is similarly given by an empirical result as $F=\mu_0 2\pi/(2R)$. Here μ_0 is the constant defined as the permeability of empty space. The determination of the field around a bar magnet is approximated by the Biot-Savard Rule. The field lines in Figure 3.3 are pictures used to represent how the magnetic force is distributed in space in Figure 3.3 below.

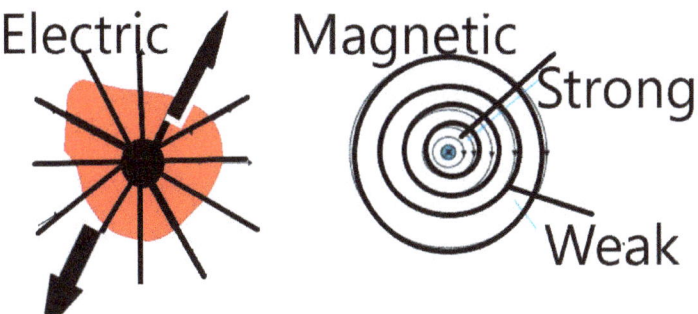

FIGURE 3.3 Gauss Laws Respectively for Electric Magnetic Fields.

The left image shows electric field lines around a stationary charge extending to empty space. The image on the right indicates magnetic field lines that wrap around a current carrying wire directed into the page. The magnetic and electric fields represent different perspectives upon the same process since motion is relative. The mathematic representation of these two situations are given by the "Gauss' Laws" which have been incorporated into a set of equations known as the Maxwell Equations. These results are expressed in Figure 3.4 below and are due to James Clerk Maxwell (1831-1979). They were published in an early form in a four-part paper between 1861 and 1862. Also, they are in his 1873 treatise on Electricity and Magnetism (references).

Name	Differential form
Gauss's law	$\nabla \cdot \mathbf{E} = \dfrac{\rho}{\varepsilon_0}$
Gauss's law for magnetism	$\nabla \cdot \mathbf{B} = 0$
Maxwell–Faraday equation (Faraday's law of induction)	$\nabla \times \mathbf{E} = -\dfrac{\partial \mathbf{B}}{\partial t}$
Ampère's circuital law (with Maxwell's correction)	$\nabla \times \mathbf{B} = \mu_0 \mathbf{J} + \mu_0 \varepsilon_0 \dfrac{\partial \mathbf{E}}{\partial t}$

FIGURE 3.4. Maxwell's Equations.

The first two equations are the Gauss Laws pictured in Figure 3.3. The third equation tells us that the partial derivative of the energy with respect to time correlates with a change in the magnetic field. The fourth equation states the reverse and that a change in the magnetic field produces a change in the electric field. The experimentally determined constants ε_0 and μ_0 give the permittivity of empty space for electric and magnetic fields respectively. The charge density on a particle is ϱ and magnetic field lines depend upon both the amount of charge and these constants.

The three dimensional nature of light follows as the two oscillating fields are orthogonal to one another and to the direction of propagation. A simple conclusion must be that any particle, planet or universe made from light must have three dimensions. Maxwell's Equations show that the speed of light in a vacuum, c_0, is related to the two magnetic constants. The experimental result is that the magnetic permeability, μ_o, and the electric vacuum permittivity, ε_0, give the speed of light as $c_o = 1/(\mu_o, \varepsilon_0)^{0.5}$.

Maxwell was aware that he lacked a mechanism for charges, fields and light transmission. The answer to light transmission that Maxwell offered involved the presence of mechanical vortices in space. Science historians report that Maxwell's hypothesis was lampooned by many in the science community. Few or none have attempted an explanation since.

The following explanation for charges and the fields they generate is therefore new. The intrinsic property of matter that we designate as charge has its source in the light that occupies the body of matter. The motion of light within the body of a particle creates a flow of virtual photons that make up the electromagnetic field. A macro world analogue of this is any strong circulation of matter such as a quasar or tornado where rotation creates a flow.

This means the electric and magnetic fields are due to the circulation of light within the body of the particle. The light making up the matter particle is spread through the body of the particle. Circulation of light pulls virtual photons from the background and pushes them into a stream. These photons represent borrowed energy from the background that is returned in a short time. A stationary particle draws photons from the source background and pushes these into a stream that returns energy to the sink. This is evident in the first Gauss Law and cartooned in Figure 3.3. A moving entity allows the stream of photons to wrap back to become the source. These two situations describe the electric and magnetic fields as diagramed in Figure 3.5.

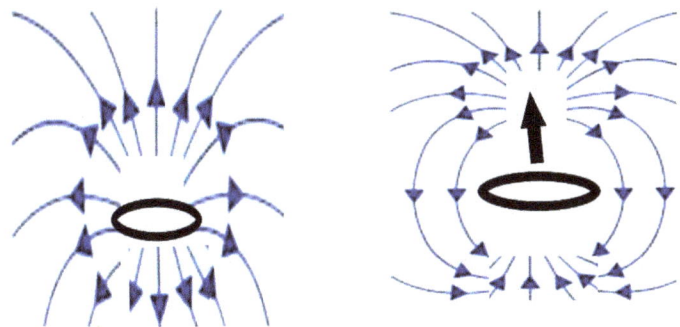

FIGURE 3.5 Stationary and Moving Particle Fields

Therefore all matter particles carry charge because circulation of light within their bodies creates a field.

The attraction or repulsion of charged matter is due to constructive or destructive interference in the fields generated by matter. These fields are created by the two charges and can be differentiated upon their respective frequencies. When like particle fields meet, then each sees a source and constructive interference follows. When destructive interaction between fields extinguishes the light between unlike particles, then each particle moves toward a sink. The attracted matter particles reorganize themselves down a path toward a radiation sink. This is modeled in Figure 3.6 below.

FIGURE 3.6 .Attraction (left) Repulsion (right)

Difference frequencies for the virtual photons emanating from charged matter can explain field based repulsion or attraction.

The two electromagnetic poles are balanced against one another indicating they are produced in a singular event. This is seen in an electron and positron pair creation event. The quarks charges are also balanced implying a creation reaction producing one top and two bottom quarks in a charge conservation pair production.

A charged particle therefore sits in its electromagnetic environment while extending its influence into the surrounding space. Light transmission through empty space depends upon two parameters that are ε_0, μ_0. These two constants of nature represent the conductivity of empty space for electric and magnetic fields respectively. The mechanism for light transmission must use matter as a source of the two fields that carry the energy. The energy in electromagnetic radiation must *form* into ephemeral matter particles that produce the oscillating fields. These virtual particles hold an electric field then move to produce a magnetic field and then decay and pass their energy to new virtual particles and repeat. The number of cycles per second gives us the frequency of the particle extinction for the energy being transferred.

The Planck Constant is a measure of the energy steps associated with light transfer and are contingent upon the matter formation rate. Increases in energy correlates with increase in the rate of virtual particle formation in light transfer.

The result of these observations is that light and matter are linked together in a recursive handshake where light produces matter and matter produces light. This boils up to us from the Plank length and below to produce our universe of matter and light. The result is that matter and light are never separate.

In summary a charge is the emergent property of matter to create a flow of virtual photons from the background environment. Electric and magnetic fields are made from this flow of light and stabilize the particle. The result is a field made from virtual photons that flow to create a field extending into the surrounding space.

SPIN

This property is like charge in the complete absence of any theoretical explanation for its emergence from matter. Spin has been described as the fundamental property that is the least understood. The physical manifestation of spin is that a particle appears to have angular momentum. Matter particles exhibit a dipole and this means that the charge is unevenly distributed across the body of the entity. This is consistent with the evolving model of matter that describes a particle as a bundle of light that is dispersed across the body of a particle. Charge has been described as due to the light inside the particle and this is spread unevenly through the body of matter. Charge distribution is a widely used concept in chemistry and physics.

The light within the particle is distributed through the volume of the particle and this is mutable. The fields that influence distribution determine the spread of charge within the body of a matter particle. The confinement field that shapes a matter particle can be affected by the application of an external field. The result of this is that a dipole can be induced for any matter particle. The dipole created by the uneven distribution of circling light is cartooned in the Figure 3.7 below.

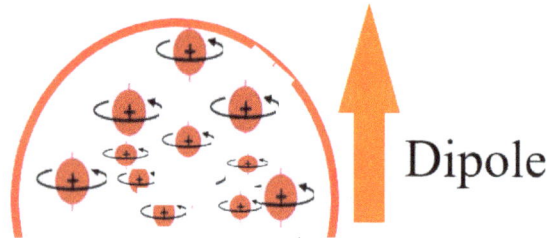

FIGURE 3.7. Distribution of Light Creating Dipole.

The macro world analogy for the dipole created by charge distribution is in that of air moving away with from the surface of the earth. The energy density inside a particle deceases as away from the center of the particle. The result of this is that a charge distribution exists within charged particles.

Spin gives us the experimental methods known as Nuclear Magnetic Resonance and Magnetic Resonance Imaging. The NMR experiment is done by applying a pulsed external magnetic field to a sample. The relaxation of absorbed energy back to the ground state indicates the magnetic environment of the sample.

SUMMARY

The three intrinsic properties described above all emerge from the nature of the light making up the body of matter. They represent basic measurable properties that can be used as input for a mathematical model. The mechanisms given for these intrinsic properties of matter evolve from the model that assigns light as the center of matter. Modeling the light inside of matter is a door into the nature of matter.

Chapter 4
EXTRINSIC PROPERTIES OF MATTER

Our journey has shown that matter is made from light. Motion is due to acceleration fields that introduce light into the matrix of matter particles. This means that the interactions between matter and acceleration fields are mechanistically available. The Standard Model has four different kinds of interactions that it labels as Fundamental Forces. These are documented in Table 4.1 below showing their relative strengths, ranges of influence and proposed intermediaries.

FORCE/INTERACTION	RELATIVE STRENGTH	RANGE IN METERS	MEDIATED BY
Strong	1	10^{-15}	Gluon/Mesins
Electromagnetic	1/137	Infinite	Photon
Weak	10^{-8}	10^{-18}	W^+, W^-, Z_0
Gravity	10^{-39}	Infinite	Graviton

TABLE 4.1 Four Standard Model Interactions.

Three of these are acceleration fields and the fourth is a decay sequence. Proposed mechanisms for the acceleration fields must show how energy is introduced into the bodies of matter particles. These necessarily take into account the different environments in which they operate and their relative strengths. The Weak Interaction is the anomalous character in Table 4.1 and has a distinct mechanism of action.

The mechanisms described for these interactions deal solely with the transfer of energy and do not deal with how matter accepts energy. The wave function collapse associated with energy transfer to particles and photons is described in the chapter that follows.

ELECTROMAGNETISM

The electromagnetic interaction is the one that the Standard Model does the best job of explaining. This interaction involves an energy transfer between matter particles and this takes place outside the nucleus. Photons are expelled by a particle upon deceleration as described above. The photon is then available to deliver energy to matter that it encounters. An example of this occurs when hot matter in the outer surface of our sun is decerated to produce a flow of photons. These travel through space to deliver energy to our planet. A simple transfer between a photon and a particle is modelled in Figure 4.1.

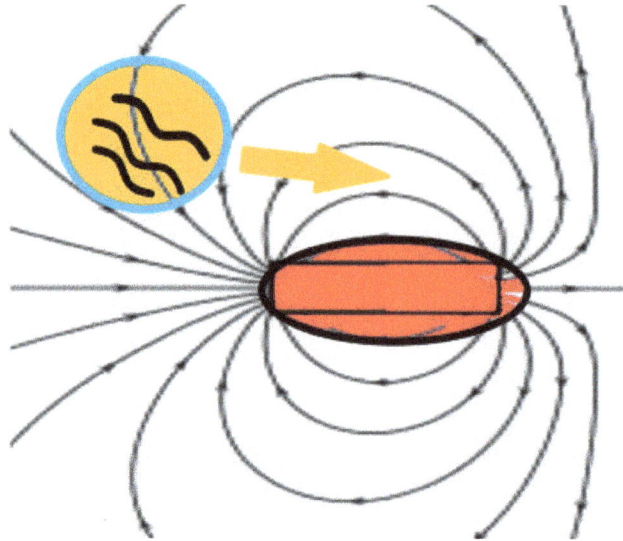

FIGURE 4.1 Photon Delivers Energy to a Particle.

Here the trajectory of the particle is changed by energy transfer as the energy of the photon is absorbed by the electromagnetic fields of matter. The wavelength of the incident light must overlap with that of the dimensions of the matter particle.

In collisions between matter particles then their respective fields overlap allowing direct energy transfer. This takes place when energy flows from hot to cold as dictated by the Second Law of Thermodynamics. This transfer is cartooned in Figure 4.2.

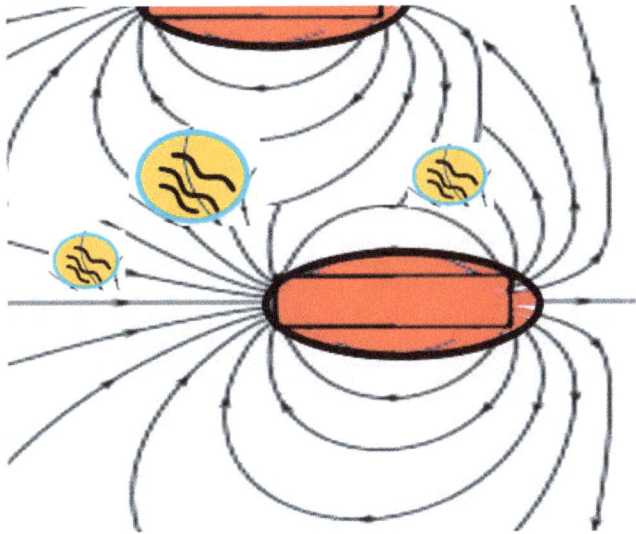

FIGURE 4.2. Particles Transfer Energy by Field Overlap.

The result is that matter upon matter energy transfer is a common mechanism for accelerating objects. This is a matter upon matter analogue of the Compton Effect that describes the acceleration of matter through the influence of light upon matter.

STRONG INTERACTION

The Strong Force is a version of the electromagnetic interaction but operates inside the nucleus. Things are different in this environment where quarks are the matter particles that play a central role. These particles are composed of light and also generate a virtual field due to circulation within the body of each particle. Quarks produce gluons in a manner parallel with that described above for particles outside the nucleus. Wave collapse also occurs for these fermions just as it does for all matter particles. The action inside the nucleus differs from that outside by being higher in energy and shorter in distances. The energy density is five orders of magnitude greater than that of the electron cloud outside the nucleus. The distances inside the nucleus are six orders of magnitude smaller than even the shortest interaction outside. The net effect of higher energy and shorter distances make the interactions take less time and changes the nature of the players.

The above description of matter is that fermions are produced by high frequency light. Inside the nucleus the wavelengths are short and a flow of virtual photons is replaced by a flow of matter particles. Here the high energy radiation forms matter particles as soon as it is pulled into existence by the light inside the quarks. Recall that matter particles carry charge and the charges need oppositely charged partners. This mean that the gluon field is composed of charged matter particles that exist in partnership with their anti-matter analogues. When these charged pairs escape the nucleus they are called mesons. These act as binding gents to hold protons and neutron together in the atom. At this level of organization up and down quarks hold opposite charges and create fields that have positive and negative interference. Inside the nucleus the charges are balanced wit +2/3 and -1/3 of a full electron charge. These two kinds of charge create three particle nuclei that are either positively charged protons or neutral neutrons. Protons and neutrons with their up and down quarks labeled with electric and color charge are depicted in Figure 4.3.

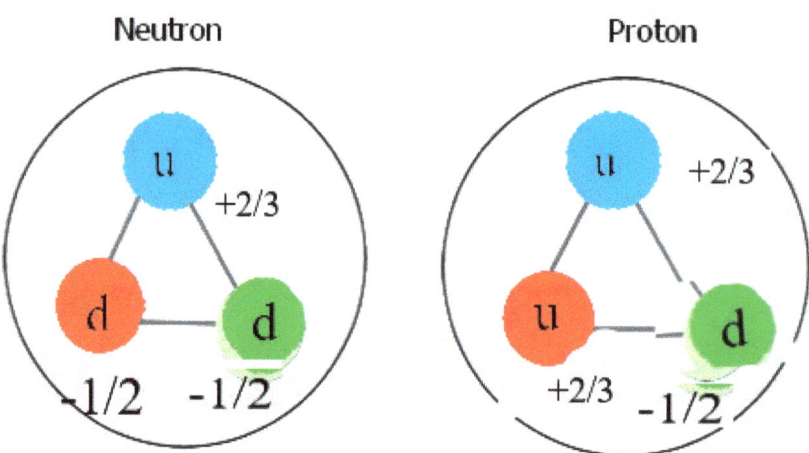

FIGURE 4.3 Quarks inside the nucleus.

The Strong Interaction mediated by these solid charged intermediates is 139 times stronger than the Electromagnetic Interaction. This interaction has the unusual bonding character that the strength increases as bound pairs are separated. This contrasts with the electromagnetic interaction outside the nucleus that falls off as the square of the distance between ions. Another difference is that outside the nucleus individual electrons can exist alone while quarks only exist inside the nucleus.

The explanation of quark binding comes from a conformational analysis of the preferred alignment of three quark systems. Inside the proton and neutron the charges on the quarks force a linear arrangement. This follows from the repulsion of the two like particles that are both attracted to a central opposite charge. Mesons and gluons act like bosons because they carry the same directional symmetry as the moving body of light in a photon.

Kinetic energy accounts for 99% of the masses of the proton and neutron. The special ability that these hadrons have for holding kinetic energy comes from their preferred conformations. This preference crates a large number of degrees of freedom for motion around the central charge as diagramed in Figure 4.4.

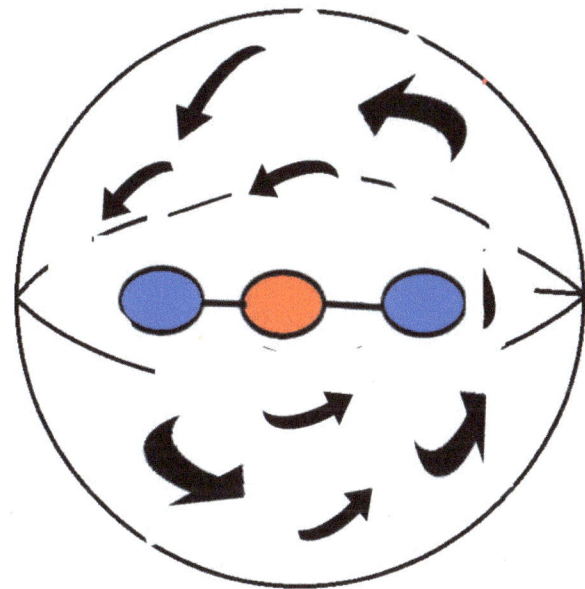

FIGURE 4.4. Quark Optimal Geometry with Multiple Degrees of Freedoms to Spin

Kinetic energy is therefore stored within the proton or neutron in the motion of their substituent quarks. The kinetic energy increases as motion pushes up the relativistic mass. The spinning action in the nucleus is responsible for most of the mass of these subatomic entities. In summary the mass of the neutrons and protons is largely due to repulsion between like charges. The resulting conformations produce optimal geometries for spinning action by the three particle systems.

When the optimal geometry in protons and neutrons is lost then this reduces quark motion. When the optimal quark orientations are lost then spinning slows and kinetic energy must escape. This deceleration requires that the quarks must expel kinetic energy as light. The flow of gluons between charged quarks becomes an avenue for kinetic energy discharge. This increases the number of gluons expelled into the space between quarks and increases the binding potential. This explains why quark binding increases when quarks are perturbed. When high energy particles collide and one of the three quarks is ejected then a quark jet is observed. This manifests itself in a bubble chamber as a stream of mesons.

Coupling between the kinetic energy and the binding energy explains the high stability of protons and neutrons. The instability of the neutron compared with the proton can be explained by the decay products. A neutron decays into a proton and an electron and this is a stable system while the proton does not have this option.

The stability of matter in our universe exists because quarks are never lost from a three particle system. The binding energy fuelled by kinetic energy is so great that a new quark and anti-quark pair are created and the three particle system survives. The result is that kinetic energy is expelled as a quark anti-quark pair. This leaves the original hadrons intact while producing a new meson. The neutron and proton are therefore stable due to their ability for converting kinetic energy into new matter. The new matter moves energy away as mesons that hold atoms together. The escaping mesons can deliver energy to the binding between the protons and neutrons in an atomic nucleus. When mesons exit outside the nucleus they decay into electrons and positrons.

The Standard Model holds that quarks also exhibit color charge that comes in three forms. The above description of quark binding means that the strength of the Strong Interaction is not due to the color charge. This is evident since mesons bear the color charge and are unstable. Mesons carry color charge and yet have lifetimes below 10^{-15} seconds. This suggests it is not the color charge that delivers strength to the strong interaction and stability to hadrons. The color charge on quarks can toggle between three settings for all the quarks in a hadron. This indicates that a magnetic field is at work. The color charge was determined to exist when it was observed that three identical quarks could be found in a nucleus. This meant the existence of an additional principal quantum number for quarks and this became color.

The color charge acts like a magnetic field that can be switched quickly. The color charge can be explained as an orientation in a magnetic field. Here quarks moving at high speeds in a circular motion produce a magnetic field. The color charge comes from alignment within the magnetic field created by the spinning motion of charged particles. This means the quarks each hold a different alignment within the field they create by their motion. The field created in this way can have a mirror image and this is seen in the color and anti color pairing. This means the color charge has six different poles. A change in the direction of a magnetic field translates into a change in the color of individual quarks. Each of the three quarks inside a nucleus must align their magnetic fields with the prevailing field.

The gluon mediator can change the color and therefore field orientation of the accepting quark. The field change is comparable to the way the earth's magnetic field can toggle between two orientations. A color charge on the quarks is a physical property that emerges from the combination of charges and motion. A computational representation of the quarks would yield information about the nature and interactions around the color charge.

In summary the Strong Interaction is a boosted form of the electromagnetic interaction. The quark conformations within the hadrons impart stability upon their systems. Energy is processed into matter as the three quark system accepts energy and converts that into matter. The stability of protons and neutrons, that are 99% kinetic energy, gives our universe its solid nature.

GRAVITY

A benefit of travel to new lands is the chance to obtain new perspectives. Gravity is a force that we have known since our ancestors traveled through the branches of a tropical forest. This acceleration field is found in our skeletons and muscles. Old explanations for this interaction are unsatisfactory. The warping of empty space in Relativity theory and the graviton in the Standard Model can be dismissed when a better explanation becomes available.

The two acceleration fields discussed above have opposite charges and function by streaming virtual photons or gluons to create fields. The gravitational interaction is only attractive and is much weaker than all other acceleration fields. The Standard Model holds that gravity is a field moderated by the proposed and never observed graviton. The idea of a graviton messenger radiating from a gravity source fails. This proposed intermediate boson would need to escape a black hole to produce gravity.

Gravity is weakly interacting and only attractive and this implies a different mechanism from the other acceleration fields. The universal nature of gravity is evident in the way it draws matter, light and dark matter. Operating outside the nucleus at infinite distances gravity is 39 orders of magnitude weaker than the strong interaction from Table 4.1. Gravity attracts at the speed of light and Newton gave us the basic equation for attraction between masses as $F=GM_1M_2/R_{12}^2$. Einstein made minor additions to Newton's equation and explained the attraction as a warping of empty space. The gravitational attraction looks like the electrostatic force between charged particles as noted above as $F=Kq_1q_2/R_{12}^2$.

Using the knowledge that matter is made from light it is possible to conclude that gravity is a light upon light interaction. The field is propagated through empty space and acts to attract light toward a source that is also light. This light upon light interaction is ultimately responsible for all other interactions because it creates matter. This force is responsible for turning high frequency light into matter. A mechanism for the light upon light interaction is therefore possible. The hypothesis here is that travelling light absorbs energy as it passes through space. The deficit of background energy in the volume of space where the light has passed then withdraws energy from its surrounding space. This motion of energy toward the light path creates a field gradient in the electromagnetic background. The proposed mechanism for light upon light attraction is cartooned in Figure 4.5 below.

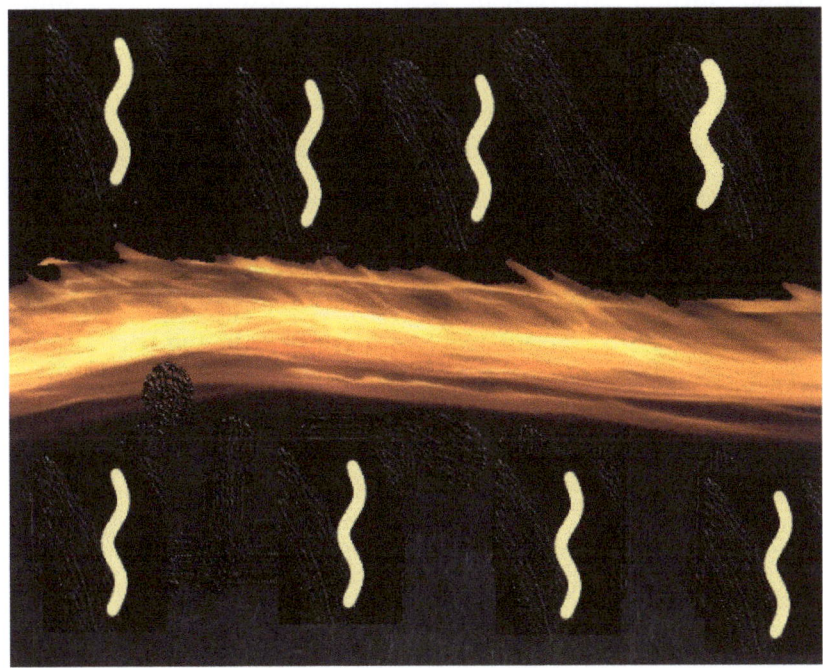

FIGURE 4.5. Light Creates a Gravity Field

The result is that matter and light are drawn to a source as cartooned in Figure 4.6.

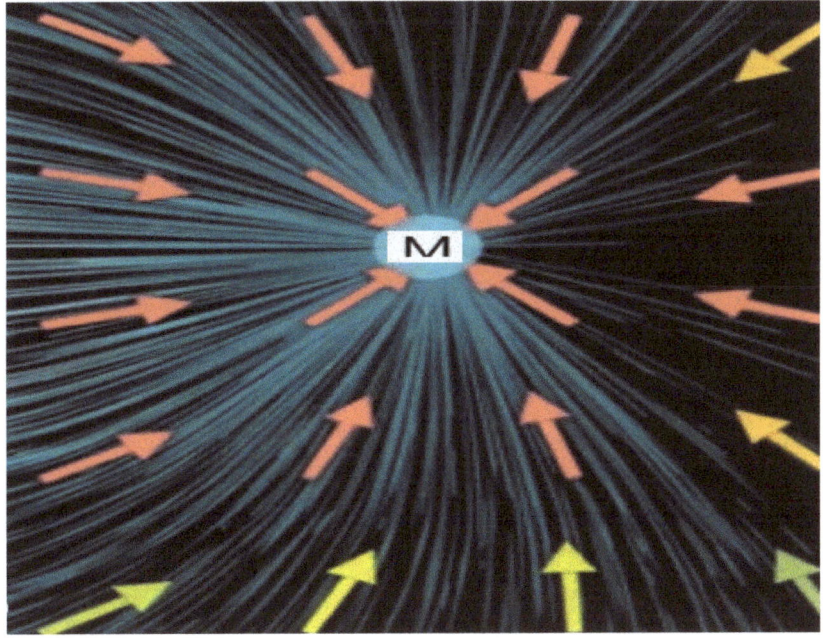

FIGURE 4.6 Gravity Field

A matter particle placed in a gravity field gains energy from the surrounding electromagnetic background. In effect the gravity source draws energy through empty space this produces a field. Matter placed in a gravity field draws energy from that field and is accelerated toward the source. Matter held in place in a gravity field, like the particles in your body, has its radiation pathway distorted. This stores energy in the particle light changing time passage and relativistic mass. In summary the gravitational interaction is the most fundamental of the four listed in Table 4.1. This interaction is mediated by empty space as a gradient in the background electromagnetic environment. The effect crosses all frequencies and gravitational attraction scales with the total amount of mass or energy. The high density of light in matter particles makes these entities the most attractive.

THE WEAK INTERACTION

The weak force is described as the most unusual of the four types of interactions listed in the Standard Model and above in Table 4.1. Fundamental properties of the Weak Interaction include that it treats matter and antimatter differently. The weak force interacts only with clockwise spinning matter and counter clockwise spinning antimatter. The Weak Interaction is asymmetric with respect to time by preferring a directional feature that unwinds but does not rewind. This weak interaction is mediated by both charged and neutral high mass carriers that are labeled W+, W- and Z bosons. These have masses as listed in Table 4.2.

FORCE NAME	NAME	MASS GEV/C²	ELECTRICCHARGE
Electroweak Spin = 1	Photon	0	0
Electroweak Spin = 1	W-	80.4	-1
Electroweak Spin = 1	W⁺	80.4	+1
Electroweak Spin = 1	Z⁰	92.187	0
Strong Spin = 1	Gluon	0	0
Neutron Spin = ½		939.57	0
Proton Spin = ½		938.28	+1

TABLE 4.2 Bosons, Protons and Neutron Masses.

The change that takes place converts a neutron into a proton, an electron and an antineutrino as shown in Figure 4.7.

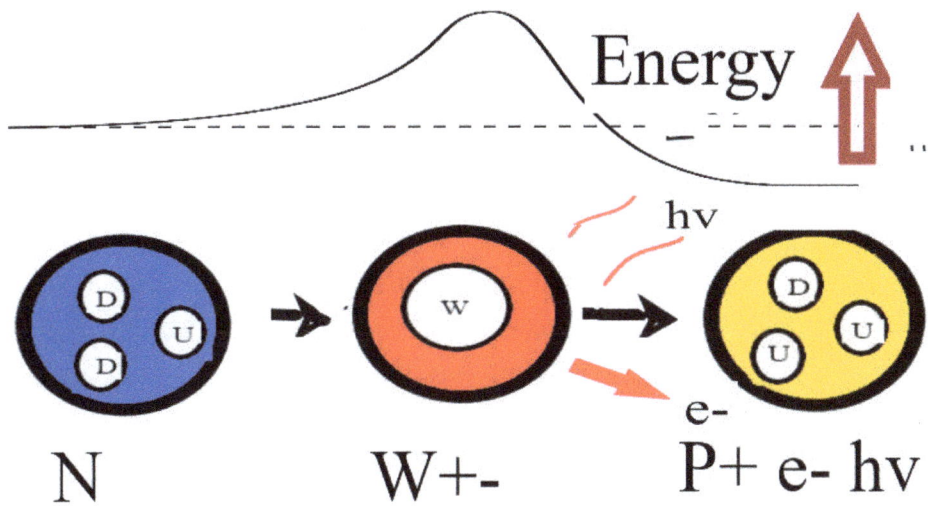

FIGURE 4.7. The Weak Reaction

The above description of the conformations and energies of the three system neutrons and protons now comes into play. This conversion requires a complete reorganization of the three particle system. Two down quarks become a single down and one up with the other up quark remaining intact. The kinetic energy of the neutron is available to become a boson intermediate with kinetic energy. The mediators then decay into a new three quark system giving off an electron and an intermediate. Charge and energy are conserved in the product of this sequence of decay reactions.

A general description of the decay process is useful at this juncture. When the moving light contained in a matter particle, or in this case a system, finds a low energy state then it collects in that state. The achievement of a lower energy state may depend upon the height of the transition energy into the new state. In the three particles system of a neutron there is ample kinetic energy available to produce a matter particle with kinetic energy. The masses of the proton, neutron and weak bosons in Table 4.2 tell us the system goes through a net mass valley. However, we know that mass represents only part of the story for these three particle systems. The infrequent nature of the weak conversion suggests the intermediate has a net energy peek.

A general description of the decay process is useful at this juncture. The case for atomic decay such as the decay of Uranium into Lead occurs as a series of alpha and beta decays. Alpha decay separates a helium atom consisting of two proton and two neutrons from the original atom. When a fundamental particle decays, then some of the light making up the particle separates from the main body of energy. This light can be ejected as pure light or as a lower energy particle. This situation is seen in the weak interaction that converts a neutron into a proton with the associated electron and antineutrino ejection. The neutron decay is energetically favorable as the charge separation moves from inside the nucleon to outside.

The difference of 1.29 GeV/C^2 between the proton and neutron is motivation for this decay reaction to occur. The importance of this transformation is that it is responsible for sunlight by mediating neutron decay in our star. Like the decay of unstable nuclei, fundamental particle decay involves separating building blocks.

Neutrinos travel at, or close to, the speed of light because they are strings of high energy light. Neutrinos do not react with matter other than through the weak interaction that allows a quark of +2/3 to change into a quark of -1/3 and an electron. This reaction can be reversed and happens by the transition through a W or Z boson. A quark and an anti quark can combine to become a W boson. The weak interaction can change a neutrino into a lepton such as an electron, a muon or a tauon. Only the top quark (173Gev/C^2) has more mass than the W bosons and so is the only particle that radiates easily by the Weak Interaction.

In summary the weak interaction is a route through which particles can interconvert one into another. The conversion that produces an electron, a proton and an antineutrino from a neutron is just one example of the plasticity of matter. Since light is the currency of matter, the conversion of any particle into any other particle becomes theoretically possible. This can be done by adjustments in light content. This treatment of the weak interaction is incomplete and a computational representation of matter is needed.

This is the juncture in our journey in which the metaphoric camels are loaded and ready to caravan. The flaps inside the camel stomachs are bloated with water and supplies are packed each in its place. The model has provided mechanisms for mass charge and spin while explaining how particles experience acceleration and move. Now that our travel vehicle is prepared we can move into the scenery that lies ahead. Our journey continues to the valleys and hills of an old friend in the guise of Quantum Mechanics where new perspectives arise.

Chapter 5
QUANTUM MECHANICS

Quantum Mechanics is a body of knowledge that deals with individual particles. A starting place for journey through this topic is the Schrödinger Equation.

SCHRÖDINGER EQUATION

The Schrödinger Equation (Edwin Schrödinger 1927) is central to quantum calculations of the total energies of fermions. The kinetic and potential energies for fermions are calculated using experimental mass and charge. The total energy is minimized by varying contributions in the equation (Etot= Kke + Epot). Electrons feel the charge interaction of a positive field E**pot,** due to the atomic nucleus that holds them. The equation has the constraint that all matter particles must have wavelike behavior. The wavelength is given by the de Broglie relation described above that combines hv and mc2. The final version of the equation is

$$H\psi\hbar = E\psi\hbar$$

Here, H, refers to the Hamiltonian representing contributes to the total energy. This is simply the Kinetic energy and the electric field interaction. The psi, ψ , refers to the wavefunction that constrains the motion to obey wavelike motion. The E represents the energy obtained by the individual contributions collected in the Hamiltonian. The equation calculates the instantaneous energy for a charged particle in a field.

A cartoon widely seen by the scientific community depicts a student pointing to the great equation and asking "Why does it work? Neils Bohr answers "Shut-up and compute". This expressed the fact that the scientific did not know why the Schrödinger Equation worked. The fact was that it did work and correctly calculated energy levels at a high degree of accuracy as seen in QED.

The answer to the question of why the Schrödinger Equation works is now possible. The model of matter developed in the first chapters of this book has an explanation for the effectiveness of this simple equation. The Schrödinger Equation treats a particle as a point charge with mass. The routine is to test the energy at a set of coordinates using the equation and making a map of electron density. The map is a probability chart indicating where the electron has a chance of being found. The light making up the electron is spread through the orbital as a function of attraction to a positive center. The kinetic energy and the position of the particle determine the spread of light within the volume of available space. The calculation works because a point charge position map is the same as the actual spread of particle light. Chemistry is based upon the spread of charge between atoms and molecules and this verifies the model. The calculated result gives the true result by serendipity rather than a correct theoretical comprehensive. In effect the probability map obtained by the equation mimics the map created by nature. The light making up the electron spreads to fill the available space.

The electron can be found according to the confinement created by attraction to a positive center. The Schrödinger Equation gives a map for electrons occupying the 1S, 2S and P x,y,z orbitals as shown in Figure 5.1.

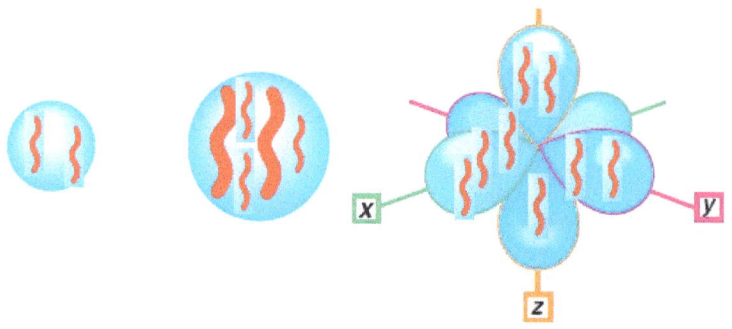

FIGURE 5.1 Maps of 1S, 2S, 2P Orbitals.

Some orbitals have lobes that are separate from one another. This is possible when the light making up the particle is spread through the volume. This is achieved by a particle that is made from building blocks that are individual units of light.

The charge on the particle is held in the light making up the particle and is spread like the source light. The kinetic contribution is dictated by the attractive field. In the absence of any field the particle has the freedom to move and will spread itself into the surrounding space as in Figure 5.2.

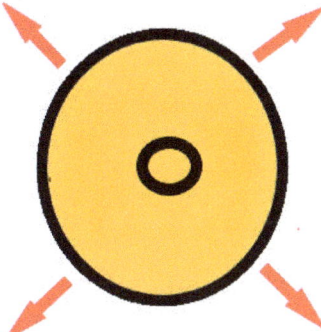

FIGURE 5.2. Electron Field Expanding.

The macro world analogue of this calculation is the measurement of the intensity of light in a room that has shadows and bright areas. The light is most intense where the field is strongest and less intense where shadows occur. The light in an accelerated electron spreads with the velocity of the moving particle. This is modeled in Figure 5.3 below.

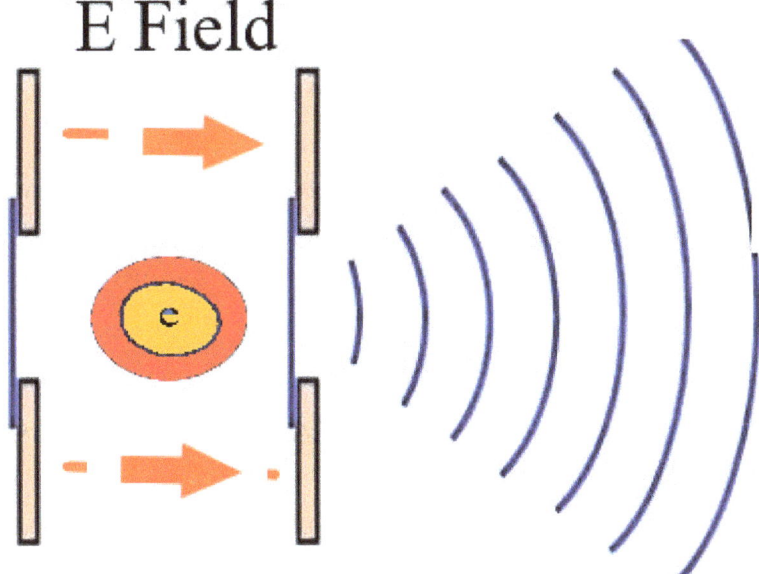

FIGURE 5.3 Electron Spreads After E Field Acceleration.

QUANTUM ENTANGLEMENT

This experimental result is a black letter fact of nature. Particles and photons both exhibit the quantum phenomenon that is known as entanglement. The experiment that proves quantum entanglement is simple in principle and difficult to complete. The experiment is depicted in Figure 5.4 below.

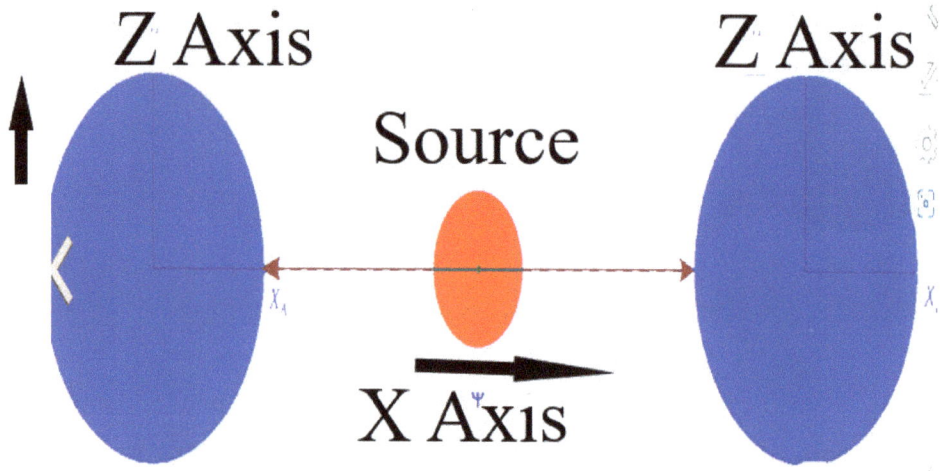

FIGURE 5.4. Experimental Determination of Entanglement

The procedure involves producing entangled photons or particles and measuring properties such as spin for the separated entangled pair. A difficulty comes from keeping the pair entangled while separating them at a distance suitable to determine faster than light action. As recently as 2015 experiment has verified that measuring a property on one entangled partner, determines that of the second. The measurement on the separated pair has to be done before light can travel between the two. This area of Quantum Mechanics dates back to the early days of the use of the Schrödinger Equation. Here computational results suggested that particles in close proximity could have their respective wave functions mixed. The argument against entanglement was made in a paper by Einstein, Podolsky and Rosen entitled, Can a Quantum-Mechanical Description of Physical Reality be Considered Complete?. The thesis of this paper was that hidden variables could be responsible for the appearance of entanglement. This was presented in opposition to the interpretation given by Neils Bohr. Here the Copenhagen Interpretation of Quantum mechanics held that wave collapse for both partners occurred simultaneously. Experimental results have demonstrated that Bohr was right. The overlap between two light bundles is depicted in Figure 5.5.

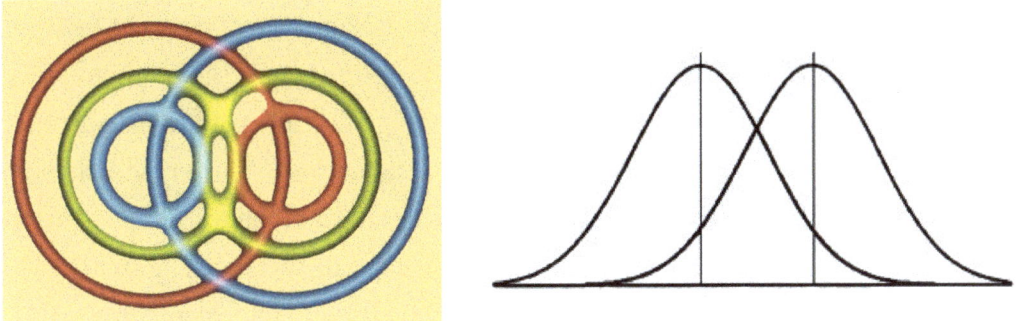

FIGURE 5.5 Overlapping Entangled Pairs Mixing.

A computational indication that mixing was possible was followed by experimental verification. However, none were able to provide a mechanism that explained this observation. Einstein famously called this "spooky action at a distance". Treatment of matter as a black box made it little wonder that entanglement appeared to be mysterious. Having at this juncture opened the black box, it is apparent that quantum entanglement is a natural product of light based entities. The conclusion is that standing waves in the electromagnetic realm are responsible for Quantum Entanglement. Standing waves are common in nature and we can observe one by moving two ends of a skipping rope up and down in unison. Standing waves can be created in air, water or light and the case for light is depicted in Figure 5.6.

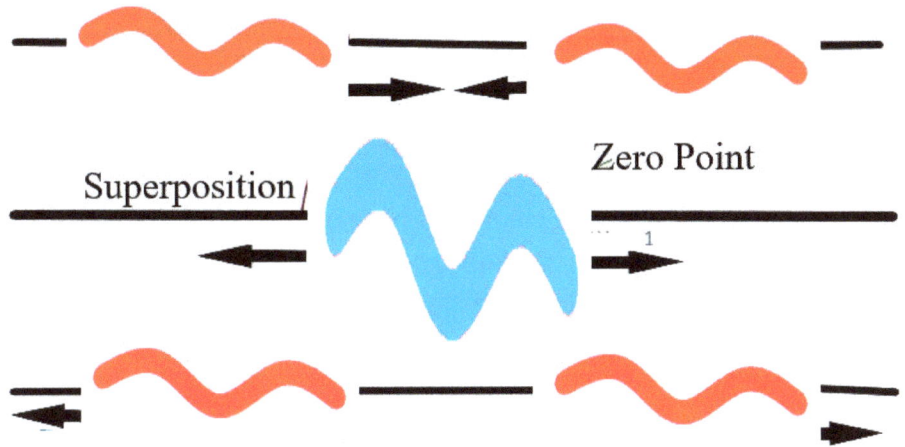

FIGURE 5.6. Standing Wave in Light.

In this figure the top line represents the two incident waves of light. The second line represents the standing wave that results from the collision of the two waves. The third line represents the two waves that recede back to the two sources. The mechanism for entanglement between particles and photons is that standing waves can exist between separated bundles of light. Mixing of the light making up any particle or photon produces a standing wave when separation occurs. The result is that bundles of light, once separated, remain connected as electromagnetic standing waves. Like an old fashioned taffy pull, strings of the sticky substance dangle between the separated lumps of material. This situation is cartooned in Figure 5.7. These standing waves can be stretched to reach across space and a collapse on either partner ends the entanglement.

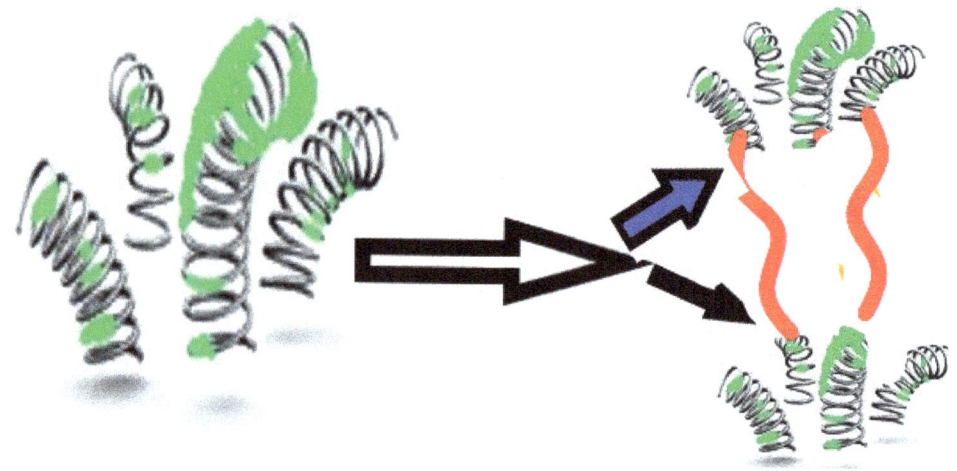

FIGURE 5.7 Entangled Standing Waves Connecting Pools of Light.

The standing wave is not an energy transfer, but a link between separated light clumps.

This means the question of non-local reality can be answered. Particles can become entangled when held in close proximity or upon creation events that produce a pair. The entangled pair is connected by standing waves created when separated sides send equal frequency signals. When a measurement is made then energy is transferred and the standing wave on one side is shut down. As soon as one side relaxes its input into the standing wave, then the entire standing wave collapses. This produces a wave collapse on the entangled twin at the same instant of the initial collapse.

The answer to the question of locality is that it exists in all cases as information does not travel faster than light. This follows because the measurement on one of the pair produces a random answer. This means that reality persists and that faster than light information transfer does not occur. Entanglement is a special case for matter or photons that have their bodies mixed. The junction between pairs is precarious and easily lost. Any energy transfer that takes place for any measurement causes collapse and instantaneous disentanglement. Hidden variables do not exist and wave collapse happens for the two entangled partners at the same time. The conclusion is that entanglement exists for all separated bodies of light. This phenomenon exists, not only between fermions and bosons, but also within the bodies of these entities.

WAVE COLLAPSE AND MEASUREMENT

A measurement requires an energy transfer from or to a matter particle or photon. The kinetic energy that is stored as light in the body of the particle is transferred. Energy transfer between electromagnetic fields obeys the Second Law of Thermodynamics. This rule states that energy moves from hot to cold when an avenue for transfer opens. In any measurement, such as in the case of the "Double Slit" experiment, an energy transfer occurs. The kinetic energy is transferred as soon as a conduit for energy transfer opens. This happens when the moving field of light encounters a transfer point at the detector plate. When fields of the particle and the sensor overlap, then energy transfer follows and a signal is recorded. The dispersed field of light that made up the moving particle or photon field is concentrated into a local volume of space.

This situation is comparable with the macro world event that we know as lightening. A matter particle that consists of a moving front of light releases its kinetic energy in a single discharge like the lightening strike. The change that occurs upon wave collapse is not described by the Schrödinger Equation. The pre-collapse state before measurement is given by the equation as a probability function of states of the matter particle. The physical result is given by experiment. The wave collapse is therefore a transition from a range of potential states to a final result.

Once a window for energy transfer opens, then virtual photons flow between over-lapping electromagnetic fields. The full load of available energy transfers at the speed of light. The wave collapse initiates the disentanglement of dispersed components and their energy is moved to the background and then reclaimed at the site of transfer. The final state selected is one with a high QED probability and the one that opened the transfer window. The result of a measurement is that it forces the energy of the dispersed particle to collect at the site of measurement. The matter particle then exists at the site of transfer and is available to spread into available space. The measurement determined by experiment gives a final state. The exchange of energy with the background is the mechanism in the wave collapse. The result delivers the full energy of the particle to the site of measurement. This situation is modeled in figure 5.8 below. In this figure the orange shape represents the moving electron field and the black shapes represent different states the electron field adopts. The collapse results in one of the states being selected and the energy from the field fed into the background and then into the final measured state. The red arrows represent energy flow from the electron field into the background and from the background into the final state

FIGURE 5.8. Wave Collapse Energy Flow.

Energy flows from dispersed contributions into the background and back from the background to the site of transfer. A particle delivers most or all kinetic energy to the sensor, while a photon can deliver its full complement of energy and cease to exist. The particle is refreshed at the site of transfer and is then ready to move again.

DOUBLE SLIT EXPERIMENT

Therefore we can now explain the Double Slit Experiment. The original experiment was done by Thomas Young as early as 1801 with sunlight as a source of light. The interference pattern made it evidence that light moved as waves.

In recent versions of the famous experiment electrons are passed through an apparatus as in Figures 5.9.

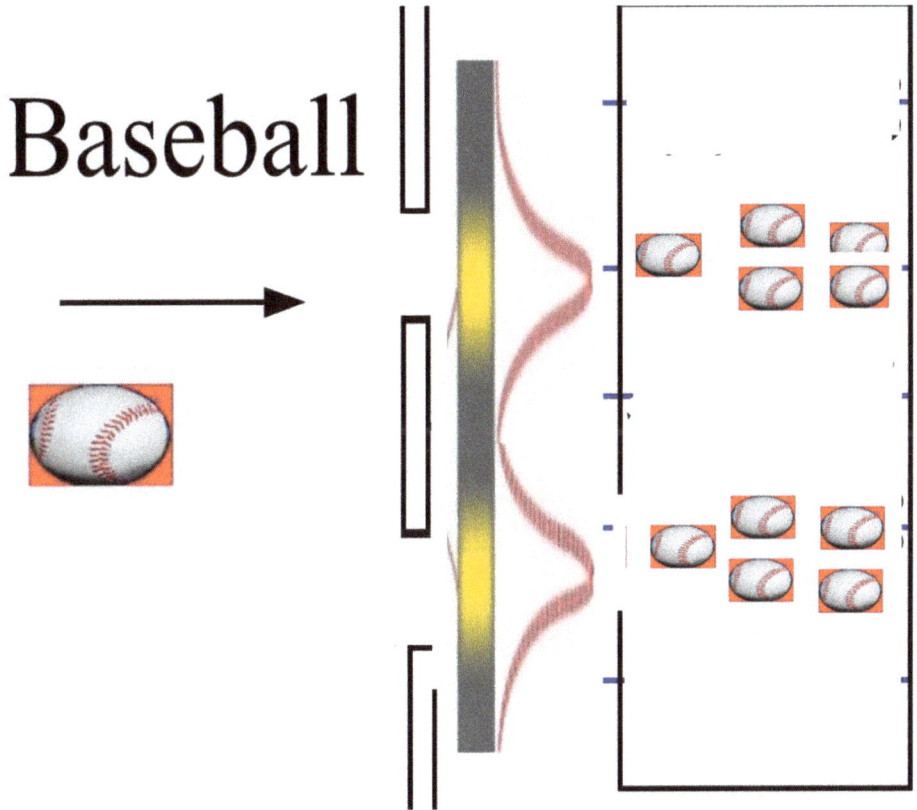

FIGURE 5.9 Double Slit with Base Balls.

The pattern in 5.9 is what happens when macro sized particles are fired at a double slit. The result is different when light is the source and then a diffraction pattern is seen. This situation persists for quantum particles and is depicted in Figure 5.10.

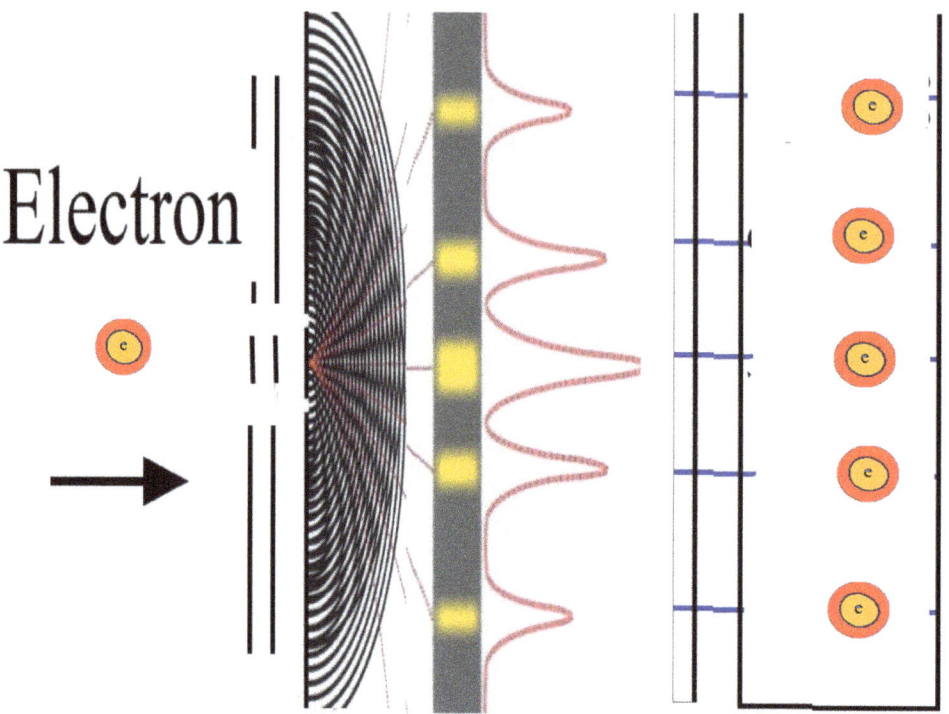

FIGURE 5.10 Double Slit With Electrons.

When the rate of electron flow is slowed to a single electron at a time, then the diffraction pattern in Figure 5.10 persists. This means the electron interferes with itself after passing through both slots at once. A particle as large as a C_{60} Bucky Ball displays a diffraction pattern. This demonstrates the wave and particle duality of both photons and particles. Here they interfere as waves and arrive at the detector as particles.

The appearance is that the electron begins its journey as a charged particle accelerated in an electric field. It then spreads out as an advancing wave of light and passes through the two slits simultaneously. The light exhibits interference as the slits act as individual sources that have peaks and valleys that overlap. When the moving light encounters a detector then it appears as a particle. The question arises around how a photon or particle that begins as a particle and arrives as a particle. The wave and particle duality of nature is expressed in the "Double Slit" experiment.

Other apparent features of Quantum Mechanics are revealed by a complete explanation of this experiment. Therefore a particle passing through the Double Slit enters as a particle that is accelerated by an electric field. The electric field is a flow of virtual photons between charged matter particles and delivers energy to the particle. The particle moves as a field at a constant velocity in the direction created by the acceleration field. The light in the particle is now free to disperse and spread through diffraction. The body of light moves at the velocity gained from the field. The light passes through the two slits that act as separate sources of light. The two sources of light create an interference pattern. The front of the moving particle reaches the detector and a transfer site is found. The detector receives energy as the entangled components of the electron are released into the background and reclaimed at the site of transfer. The completion of the transfer leaves the particle refreshed at the position of its energy transfer. The energy transferred to the detector records the position of the strike. The sequence is diagrammed in the Figure 5.11 below.

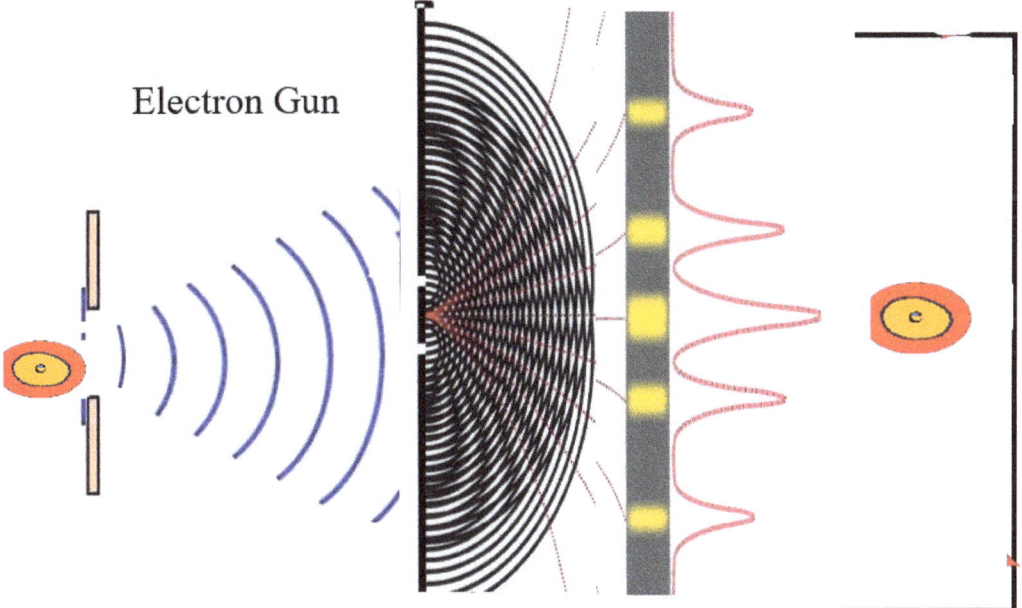

FIGURE 5.11. Stepwise Explanation of the Double Slit Experiment.

A version of the "double Slit" called the "Which Way" sets a light to test which slit the particle passed in its trip to the sensor. When the light photons interact with the electrons then the interference pattern on the screen is lost. This is explained by noting that when a photon of light interacts with the moving electron then it causes a wave function collapse.

When a photon interacts with an electron then energy transfer to the particle collapses the wave of the moving body of light. In the "Which Way" experiment the electron has its wave collapsed before reaching the detector. This breaks up the refraction pattern and explains the result observed that interference is lost.

QUANTUM ELECTRO DYNAMICS

Quantum Electro Dynamics (QED) is a computational method based upon the Schrödinger Equation. This creative use of the equation yields accurate results and is used to study particle physics. The method gets the label Quantum Field Theory. QED treats matter and light as particles and plays a marble game of collisions using energy as a metric. This accounting describes the interactions between matter as particles and light as photons. QED was developed by Feynman, Schwinger and Tomonaga who were awarded the Nobel Prize for QED in 1965. They used the energies from the Schrödinger Equation to track interactions. Their results are the best example of computation in Quantum theory. Richard Feynman (1918-1988) described this in his book QED. The Strange Theory of Light and Matter (references)

The inspiration for QED was based upon the light that takes all possible routes to arrive at any point in space. This principle was used in describing the interactions between light and matter. Each different path the photon takes has a mathematical contribution to the net intensity at the point of measurement. The idea is to use the energy from all contributions just as the intensity of light is due to the sum of contributions. Feynman developed a convention now called Feynman Diagrams that portray the exchange of photons between charged particles. His system uses a wavy line for the photon, a straight line for the electron. Different possible states of an electron and photon interaction as Feynman diagrams are shown in Figure 5.12.

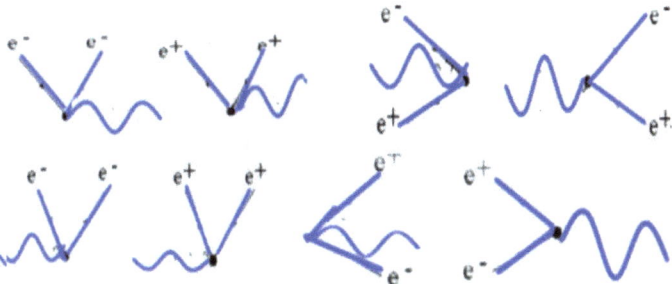

FIGURE 5.12. Feynman Electron Possible States.

The simple idea is that a photon that moves from one place and time to another place and time can take many possible routes. In QED this means all potential intermediate states contribute to the total energy and these energies are calculated using the Schrödinger Equation. The procedure adds the energies from intermediate states into that of the total energy. The best theoretical calculations of the anomalous magnetic dipole moment of the electron have a precision better than a part in a billion. This accuracy requires QED diagrams with up to four loops adding the contributions from each possible state. The sum of contributions to the whole gives the total energy of the interaction. The authors of QED were not able to explain why adding contributions into the final product gave the correct result. The answer to this is now available.

The Rosetta stone was a piece of black basalt and was a key that unlocked Egyptian Hieroglyphics. There were three inscriptions imprinted upon the ancient slab. One version of the writing was in Ancient Greek that was known and understood. The other two copies of the message were in two forms of Egyptian Hieroglyphics. Triangulation between the known and the unknown languages opened the code and produced the ability to read the ancient text. The experimental result in the Double Slit is explained by the wave collapse mechanism given above. The computational result in QRD is also explained by the same mechanism. Here a collapse combines energy from all states of the particle that exist in the moving field. The Rosetta stone of Quantum Mechanics is expressed in Figure 5.13 below.

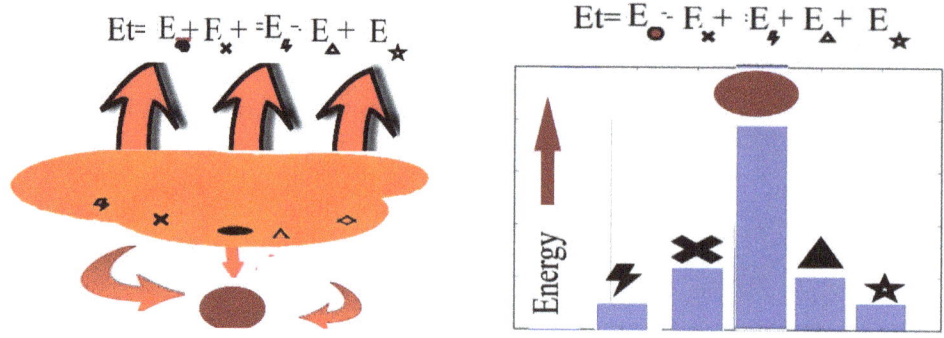

FIGURE 5.13. Rosetta Stone of Quantum Mechanics (Wave Collapse Double Slit Left, QED Right).

The summary is that a moving electron is a body of light spread through available space. The energy of the particle is dispersed among the possible states that the electron can adopt such as spin up or spin down. When energy transfer is initiated the wave collapse occurs. The energy from the different possible states is traded through the background and delivered to the final state. The accuracy of QED calculations relies upon the addition of energies from all possible states. The accurate results of QED support the wave collapse mechanism in Figure 5.8. In effect QED is a computational representation of a wave function collapse.

Richard Feynman thought QED was the crown jewel of Quantum Mechanics. He also thought that a full description of the "Double Slit Experiment" was a key to understanding Quantum Physics. The triangulation between the computational and experimental exposes Quantum Mechanics like a rat on the dissecting table. We can now use the Rosetta Stone of Quantum Mechanics to understand all relevant observations.

MISCELLANEOUS QUANTUM PHENOMENA

Therefore a quick tour of some Quantum Mechanics issues is in order. The action in the Double Slit tells us that matter moves as light but exchanges energy locally like a discreet particle. This explains the apparent dual identity of particles and photons. The Heisenberg Uncertainty Principle states that position and velocity cannot be measured simultaneously. Measurement of position requires a wave function collapse that measures position. However subtle the test for position may be, the result is a loss of kinetic energy and a change in velocity. This means a precise determination of position produces an imprecise velocity. These two are fundamentally different and therefore cannot be measure simultaneously.

The perception from the Schrödinger Equation is that a particle can exist in more than one place at a time. The above description shows that a moving particle field adopts many states that ghost one another. The A question brought up by Neils Bohr was the problem of electrons passing between orbitals without existing in the intervening space. The lesson from the "Double Slit" is that the energy of matter particles can be moved into and then back from the background. In this way particles can bypass space by being absorbed into the background in one position followed by deposition in a different position.

Quantum tunneling is a phenomenon that is responsible for sunlight. This involves the collision between protons where the two overcome repulsions and join to become a helium nucleus. The ability of matter particles to cross a barrier is explained by their energy moving into and out from the background.

Quantum Field Theory is one of the modern answers to the nature of matter. This idea was first developed by Paul Dirac to describe the interactions between particles. He developed this as a Quantum Field Theory where points in empty space were represented as harmonic oscillators storing electromagnetic energy. This interpretation has spawned a number of daughter theories that hold quantum fields as rudimentary to the properties of matter. The Quantum Field theory of matter claims that matter particles are excitations in fields specific to each particle. The only evidence of the field is the existence of the particle. The idea does not explain the source of a field that must exist before excitation. The above description of the "Double Slit" describes fermions and photons as the source of their respective fields. The idea of quantum field theory is not fully wrong, but is incomplete and does not lead to explanation. This is evident when we ask what phenomena does this idea explain and the answer is a null set.

Quantum Chromo Dynamics is the subatomic analogue of QED. This method attained its present form in the 1975 work by H. David Politzer, Sidney Coleman, David Gross and Frank Wilczek. The results of QCD calculations do not achieve the degree of accuracy achieved by QED outside the nucleus. The reason given by

QCD apologists for its inaccuracies is that gluons carry the color charge and can reproduce themselves. Their idea is that gluons can emit and absorb more gluons and that this makes their binding energies difficult to calculate. A more sensible explanation may lie in the misunderstanding around the binding energy of the gluons. The answer given above is that gluons use kinetic energy in bonding and do not use color.

The conclusion for this chapter is that Quantum Mechanics is comprehensible. The Rosetta Stone for this area of science tells us that we can explain other aspects of the quantum world. Success in area this is due to the modeling achieved by the previous chapters and is momentum for the next topics.

Chapter 6
RELATIVITY

Our journey brings us to Relativity that deals with matter in motion. This is a subject that we are well prepared for because motion is one way that matter reveals it nature. The description of matter in motion has revealed the root equation of Relativity is the **y** value.

$$\gamma = \frac{1}{\sqrt{1 - v^2/c^2}}$$

The observer of matter measures that time, mass and apparent length are affected by motion as described by **y**. The mechanisms for these have been described under the paradigm of the model for fundamental matter particles. This means exposure of this subject is already largely complete.

The fundamentals of Relativity are expressed in two postulates. These contend that light has a finite velocity and that all observers measure the same value. Galileo Galilei (1554-1642) first observed that all non-accelerated frames of reference were equivalent. The postulate became that *"the laws of motion are the same in all inertial frames of reference"*. When Galileo was alive the speed of light was assumed to be infinite. Due in part to the influence of Galileo, Ole Roemer determined in 1676 that light has a finite velocity. He achieved this by observing the moons of Jupiter. Maxwell's mathematical equations confirmed a constant speed for light. Once a speed limit was determined for light then the Galileon Transformations listed in Table 1.1 required modification. The answer came as the Lorentz Transformations described above.

Special Relativity was developed by a number of scientists including Hedrik Lorentz and Henri Poincaré. The 1904 publication by Hedrik Lorentz described the mathematics using the transformations that took his name. The link between matter and energy was made evident in the $E=MC^2$ relation that was given in 1902 by Poincare (references). This equation was noted in the 1905 German language publication by Albert Einstein that reads in English On The Electrodynamics Of Moving Bodies. (References). The Einstein footnote that first mentioned this equation gave it as $M=E/C^2$. His paper summarized the advances made upon Newton's previous work by numerous individuals. Einstein realized the mass and energy equation derived by Poincare could be used in a correction to Newton's gravity equation. Energy could now be shown to be gravitational due to its relation to mass. A question that should be asked is why it took so long for physics to go from Roemer's discovery of light speed to Relativity.

Einstein's correction to Newtonian gravity was based upon the energy to mass correction that came out of the Lorentz Transformations. The energy to mass conversion could be most easily seen in high energy environments such as near the sun. The new gravity calculation correctly predicted the bending of light near our sun and the orbit of Mercury.

Einstein discussed his ideas of a matrix representation for a gravity field with the mathematician David Hilbert (1862-1843). Einstein discussed this with Hilbert during a visit to the University of Gottingen in the summer of 1915. Hilbert independently found and published the same equations used by Einstein. Hilbert's description of the gravitational attraction was expressed in a 4 by 4 matrix called the Einstein Field Equations. These are also of the same form as the electrostatic interaction between charged particles.

General Relativity $G_{\mu\nu} = (8\pi G/c^4)\, T_{\mu\nu}$

Newton's Gravity $F=GM_1M_2/R_{12}^{\,2}$

Electrostatic Force $F=k(q_1q_2)/r_{12}^{\,2}$

The consistency between these equations is a strong hint that gravity is a field effect. This is in conflict with the Einstein interpretation of gravity as a warping of empty space.

The underlying nature of Relativity becomes evident under the new paradigm that matter is made of light. The radiation that makes up the fundamental particle of matter is responsible for the effects observed in Relativity. Mass increases are tied with the motion of matter as the light inserted into a particle adds resistance to motion. Time dilation is connected with the longer light path inside a moving particle.

The 1895 novella by H.G.Wells entitled The Time Machine described time as a dimension like the three dimensions of space (references). The idea was stated by the character of the time traveler as,

> Any real body must have extension in four directions: it must have Length, Breadth, Thickness, and—Duration.... There are really four dimensions, three which we call the three planes of Space, and a fourth, Time.

There is irony here as the book was a warning against social classes. The literary mechanism used to explain time travel became the Space-Time idea. This concept was a part of the cultural background when advanced as a scientific hypothesis in 1907 by Einstein and Herman Minkowski (1864-1909).

This means that the Einstein description of gravity as a distortion of empty space came from a scientific fantasy novel. Minkowski had recently completed a doctorate in multidimensional geometry and this influenced his interpretation of Special Relativity. He used a mathematical relationship between motion through space and time to formulate a four dimensional geometric entity labeled Space-time given by this simple equation.

$$(\Delta S^2 = c^2 \Delta t^2 - (\Delta X^2 - \Delta Y^2 - \Delta Z^2)).$$

This mixing of time and space leads to the belief that both are mutable. The x,y and z components in the Minkowski equation refer to the shape of the gravity field. The $\Delta W = (\Delta X^2 - \Delta Y^2 - \Delta Z^2)$ term is depicted in Figure 6.1.

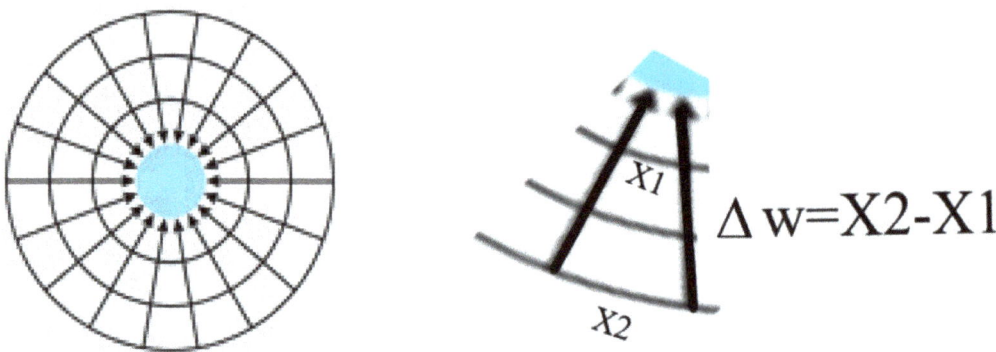

FIGURE 6.1. The Gravity Field and ΔW

The field equations treat gravity like a field and the space-time equations measure the warping of space. The alternate and mathematically equivalent answer is that empty space is not warped but that gravity is a field. The shape of the gravity field is dictated by the source of mass creating the field. Large amounts of mass organize themselves into the most efficient spherical shape. Time has been shown to pass at a variable rate that depends upon the motion of particles. This tells us that time is the only component of the Minkowski space-time equation that is mutable.

The mathematical equivalence between the two interpretations of gravity as a field or a warping of empty space concerned Einstein. He was initially opposed to the Minkowski interpretation and finally accepted this idea. In fact Einstein spent his later years trying to represent the electromagnetic interaction as a warping of space. This project would have verified the bending of empty space but was a failure.

A mechanism describing an interaction between matter/energy and empty space has never been proposed. What experiments on empty space can we use to measure anything about empty space? The low temperatures near black holes and in the laboratory rule out reaching zero degrees Kelvin. The temperature of empty space never reaches absolute zero and the result is that space is never empty. This means that a theory based upon the warping of something that does not exist by an unknown mechanism should be discarded. Based upon this simple fact we can conclude that gravity is a field effect. It is therefore possible to interact with that field and to change the gravity field. This is the mechanism used for the electromagnetic field that we read from and write to. A solution is therefore available to resolve the mathematical equivalence between warped space and a gravity field. This can be done by inventing the gravity transistor that can influence a gravity field. The gravity transistor requires a mechanism for interacting with the gravity field. The part of the field that is most interesting is the interaction range due to matter. This suggests very high frequency radiation in the background electromagnetic field should be a target.

The idea that matter bends space and space tells matter how to move is an affront to Accum's Razor. The harmonic oscillation that is the nature of light is never fully silent and the light that occupies space can be affected. The Space-Time hypothesis that treats gravitational attraction as the folding of empty space can be falsified. The determination of a field effect as the source of gravity would falsify the warping of empty space hypothesis.

The idea that gravity is a warping of empty space has essentially blocked research into gravity. The original determination of the gravimetric constant, G, was accomplished by Henry Cavendish in 1798. Measurements have improved our value of G by adding significant figures to the $(6.674 +-0.012) \times 10^{-11}\, N\, m^2/kg^2$. This means blocking a gravity field by flooding the spectrum with interference would change the apparent value of this constant of nature. Modern transportation uses environment destroying combustion to move objects. A local field damping technology that could change G even slightly would revolutionize modern engineering. The social change would rival those of the invention of the electric motor or the combustion engine. Energy could be extracted directly from a gravity field and planets could be moved in a terra forming capacity. A local influence upon G requires knowledge of the field and the ability to interface with that field.

Chapter 7

TIME

The discovery that time was not a constant and immutable fact of nature blew the minds of the early quantum physicists. This unexpected property of time fell directly from the Lorentz Transformations that followed the discovery of light speed. Visual inspection of the Lorentz y relation makes it evident that time dilation depends upon the velocity of an object. It is arguable that the discovery of the mutability of time lead to the acceptance of the warping of empty space. Many descriptions of time have been made without any mechanism for time dilation. Treating matter like a black box inhibits any chance of explaining the nature of time.

The evidence tells us that time is an emergent property of matter. Time dilation is described above as the elongation of the light path for matter particles due to motion. The Lorentz Transformations and the Double Slit Experiment have been described as experimental evidence for the motion of light within matter. The longer radiation path within the body of the matter particle causes time passage to slow for that particle and its interactions. Internal and external time exist for individual and for ensembles of matter particles. Like other emergent properties, a critical number of units are required. Wetness is an example of an emergent property and a surface is wet when many water molecules adhere to it. Time emerges when a critical number of matter particles are moving relative to one another.

INTRINSIC TIME:

The intrinsic component of time deals with the light inside the particle. The moving body of light sweeps a circle or spiral dictated by the acceleration field. A longer pathway for the light within the particle translates into a slowing of time as described above due to y. This deals with the internal motion that allows the units of light inside the particle to re-organize. Unstable particles decay into daughters when allowed the time for their light to separate. Motion for a particle like the muon slows the internal time and delays the decay process. When a statistically relevant number of iterations are reached, then a muon has a greater probability of experiencing decay.

EXTRINSIC TIME:

The extrinsic part of time is concerned with the interaction between matter particles. Motion again creates a longer route for the messenger light that travels between particles. The motion of any ensemble of matter will have the external contribution to time ruled by the y value.

Aging of matter happens when a critical number of units move through intrinsic and extrinsic time. The motion is created by energy flow as described above where energy flows from hot to cold and objects translate. The motion of any ensemble of units causes the entropy of a system to increase. The system exhibits aging as the combination of internal and external time moves the particles. The large number of units in our universe makes the chance of a recurring loop deeply improbable.

The speed of light is at the root of the rate of time passage. The result of this is that time and aging of a group are both properties of matter. These are emergent from the nature of and are affected by the motion of matter through space.

The question rises regarding where does time exists? The answer is that it exists only in the presence of matter. This means that empty space or a universe of light photons does not experience time. Consider a place in space beyond our current matter universe where matter simply does not exist with the exception f virtual particles that flash and are gone. In that place, with no matter particles, time does not exist. The ability to measure the passage of time requires a clock. Einstein's thought experiment using a light clock required a mirror to reflect the light rays. There are some places where it can be ambiguous to ask whether time exists or not. Inside a quark that exchanges energy in a constant flow, aging may not occur. Does a particle age if it can dissolve its energy into the background and reappear in a new position? Time does not appear to exist at the Planck scale at 10^{-34} Meters and it seems unlikely that we can measure time at the 10^{100} meters scale.

A number of parables were generated by Einstein to demonstrate the Lorentz Transforms. One example of this is the time dilation tale that is known as, "The Twin Paradox". Here twins are separated and one twin travels in a fast rocket and one stays on earth. When the travelling twin returns to earth the two have experienced time differently. As time dilation predicts, the traveling twin is younger than his brother.

The Twin Paradox can be described from the perspective of a fundamental particle. The starting place for the journey is in a gravity well sitting on the surface of the earth. The light pathways in the twin particles are both stretched along the direction of the gravity field. When the ship departs earth and accelerates into space the travelling twin has its pathway stretched in the direction of motion. At the same time the path is relaxed along the direction of the gravity field. Cruising speed is reached and the light pathways remain in a constant but stretched state. Time moves more slowly for the travelling twin as it stores kinetic energy in its light matrix. The change in direction required for the ship to return home means the pathway is again stretched. This can be in the form of a full relaxation followed by a repeated stretching of the light path in the particle. Time again runs slowly as the ship cruises back to earth and the stretched pathway returns to match that in the twin that stayed home. During the trip energy was exchanged between the travelling twin and its environment as energy was absorbed and emitted. The net effect is that the travelling twin has aged more slowly than his earth twin.

In summary time dilation is caused by acceleration fields acting upon matter particles. Each field causes elongation of all radiation pathways to effect time passage. The difference between two instances of time is a difference in position and state for the ensemble of particles. The originators of Quantum Mechanics treated matter particles as black boxes and were unsettled by time dilation. This may have lead to Einstein's final acceptance of the idea that gravity was a warping of space.

Chapter 8
COSMOLOGY

The theory of matter outlined in the previous chapters brings a new perspective to cosmological observations. Modeling individual particles is a starting place for modeling large amounts of matter found in planets, stars, galaxies and universes. The necessary step is to find a place where light can be blue shifted to produce high energy matter. That place is at the center of a black hole. The information gleaned from particle collisions tells us that high energy particles are formed in high energy fields. The highest energy particles decay into lower energy matter. The decay process noted above produces daughter particles that become stable protons, neutrons and electrons.

Black holes are made from matter and light under appropriate compression. Different sized stars go supernova and leave remnants that depend upon the initial star mass. This can provide us with hints about what is inside the black hole. The stratification of material in our planet and other bodies in space indicates how matter stacks under pressure and heat. The supernova is the spectacular explosion of a star that has run out of its fuel. The equilibrium between heat expansion and gravity contraction is lost and gravity takes over. This accelerates matter toward the center of the gravity well. That material either joins a solid core that has already formed or reflects back from the hot surface into space. The expulsion of hot matter into space is one of the most spectacular displays in our universe and many supernova pictures are available.

The mass of the star before supernova explosion dictates which cosmic body will be left after the upheaval ends. The ashes of burnt out stars are black holes, neutron stars or white dwarf stars. In the case of a neutron star, the pressure at the center forces electrons into the protons to produce neutrons. This is achieved through a reverse weak interaction. The masses, sizes and densities for earth, our star, neutron stars and black holes are in Table 8.1.

COSMIC BODY	MASS KG	DIAMETER KM	DENSITY KG/M³
Earth	6.2×10^{24}	12,756	5.5
Our Sun	2.0×10^{30}	1,390,000	1.4
Neutron Star	1.4 to 2.1 Sun's Mass	20	3.7×10^{17} to 5.9×10^{17}
Black Hole1	Our sun	6	*Greater $> 5.9 \times 10^{17}$
Black Hole 2	3.6 million suns	20,000,000	*Greater $> 1.0 \times 10^{20}$

TABLE 8.1 Masses and Densities of Cosmic Bodies. * estimated.

BLACK HOLE PROPERTIES

A star with around eight or ten times the mass of our star will collapse to a neutron star and stars above that mass collapse into black holes. The neutron star can show us the nature of the material at the center of a black hole. The black holes with masses only slightly larger than neutron stars must have centers much like the neutron star itself. Our own star will run out of fuel and expand into a red giant before relaxing back to become a white dwarf star. A star that eventually becomes a neutron star experiences a similar process to the one our sun will know. As in table 8.1 above the density of this material increases by many orders of magnitude in going from a planet to a neutron star. The fusion cycle inside stars eventually arrives at an end which for large stars produces an iron core. Elements higher in the periodic table than iron are only produced in supernova explosions. When the heat due to fusion ends then a rush of matter onto the core follows. This converts gravitational potential into kinetic and then into thermal energy. The center of the collapsing star is crushed under new pressure and heat due to the material raining onto the growing core.

The neutron star can be used as a template in our search for the nature of the center of a black hole. The neutron star is composed of pure neutronium material that has the density of a nucleus and is made of neutrons that are crushed together by gravity. The supernova explosion of a star that has a solar mass between 10 and 20 times that of our sun's mass produces the neutron star. Listed in Table 8.1 above, these have 1.4 to 2.1 times that of our Sun's Mass. Larger stars will produce black holes that can vary in size up to many million times the mass of our star. The center of a relatively small black hole will have a center that is only slightly denser than a neutron star. The interior of a neutron star could be promoted to strange matter if the up quarks are promoted to strange quarks. Weber et al (references) proposed the existence of a strange star somewhere in the universe as a result of a star collapse. The correct amount of energy could promote the up and down quarks into charm and strange quarks creating a strange star. Using this as an example it is easy to conjecture about the make-up of the material at the center of a black hole.

A solid core must exist at the center of each black hole with the largest most exotic at the center of the largest gravity wells. Super massive black holes have properties that distinguish them from their lower-mass analogues. Hawking radiation is more intense for small versus large black holes. This is consistent with the gradient of the gravity field around a small sphere over that of a larger example. The light inside a small black hole must have a path that is more curved than the case for larger spheres. Using an atomic term we can say that a large black hole is more polarisable than a small black hole. Communication inside the black hole is by definition not faster than the speed of light. A large black hole can therefore accommodate vibrations and oscillations unavailable to smaller black holes. The number and depth of layers in the core of a massive black hole increase with the mass of the black hole. The average density of a SMBH (defined as the mass of the black hole divided by the volume within its Schwarzschild radius) can be less than the density of water. This is because the Schwarzschild radius is directly proportional to the black hole mass by the relation $R = 2GM/C^2$. Small and large black holes differ in mass and energy content. Recall that any amount of mass or energy can be converted into a black hole by applying a critical amount of compression. As shown in Table 8.1 a black hole made from compressing our sun would have a diameter of 6 km. The estimate for the time required to dissipate a large black hole through Hawking Radiation is 10^{100} years.

Larger bodies therefore have greater pressures and temperatures at their centers. The use of space increases in value as energy is stored at maximum efficiency. Matter holds energy as $E=MC^2$ and as kinetic energy. The quark tables in Chapter 3 show the ascending masses of quarks. This follows from the assumption that the up and top quarks take about the same volume in space. The 173.1 Gev/C^2 of the top quark is much larger than the 0.002 Gev/C^2 of the up quark. This means that the volume taken up by one top quark will be 86,500 times more efficient than that taken up by an up quark. These particles have similar extensions in three dimensions because higher order particles contain higher frequency radiation. The top quark is certainly smaller than an electron and possibly smaller than an up quark. When all the matter in a neutron star is converted from up quark neutrons into top-bottom hadrons then energy is absorbed. The conversion means the required volume of space would be reduced by orders of magnitude.

The estimate is that our universe has 1.0 x 10^{53} kg of normal matter that is 4% of the total in our universe. Here 24% is assigned to dark matter and the 71% remaining is dark energy. Relativistic increases in mass due to acceleration fields are always positive. Therefore an example using 10^{53} kg as a minimum mass would be in the range of interest. The example is shown in Table 8.2 where the mass, density volume and radii of spheres are documented for the 10^{53} kg example in different states.

METRIC OF BLACK HOLE	BLACK HOLE 10^{53} SWARSZCHILD RADIUS	BLACK HOLE 10^{53} AS NEUTRONIUM CORE	BLACK HOLE 10^{53} AS TOP-ANTI-TOP MESON HADRON PLASMA
Volume Vol= Mass/Density	5.58x10^{78} M^3	1.7x10^{35} M^3	5.1x10^{32} M^3
Mass kg	1x10^{53}	1x10^{53}	1x10^{53}
Density kg/ M^3	1.792x10^{-26}	3.7 × 10^{17} to 5.9×10^{17}	5.9×10^{17}x86,000= 5.1x10^{22}
Radius M	2.33 x 10^{28}	5.53x10^{11}	1.83 × 10^{10}
Radii Ratio	4.2x10^{16}	30.2	1

TABLE 8.2 Volume, Density and Radii for Three States of 10^{53} kg.

The volume jump from the Swarszchild Radius of a black hole to the radius of a Neutronium Core is 4.2×10^{16} M. This shows that a black hole is made of a solid core surrounded by dispersed matter and radiation. Most of the volume is low density material with a growing core at the center. A change in the radius of a solid core by a factor of 30 occurs in the promotion of a neutronium core to a top anti-top-quark plasma. The change in the volume of space in this example is 1.699×10^{35} M³. The radius of a sphere of this volume is $R=3.7 \times 10^{11}$ M. This means a volume having a radius of 3.7×10^{11} M becomes available. The collapse of material into that volume would follow the promotion of up into top quarks.

Modeling the supernova collapse of material into available space is possible. In essence heat and pressure increase at the core due to accretion of matter and light by the black hole. The resulting promotion of up to top quarks is followed by an energy surge at the core. Here gravity delivers matter and light as kinetic energy and this is converted to thermal energy. The energy can be calculated for the change in volume and associated crash of matter onto a central core.

This means that we take either a half or a fifth of the total mass and suspend it at the specified distance from the center. The resulting gravitational field strengths and potential energies are tabulated. These are compared with the rest energy of the matter in the highly energetic top anti-top plasma.

The energies of the example collapses are documented in Table 8.3 below. There the two collapses represent the promotion of either 20 or 50 percent of matter to a higher order. Large amounts of potential energy are released onto the core by a collapse as described.

1.0 X 10⁵³ KG OF MATTER	RADIUS CHANGE FROM S1 TO S2 IN METERS	ENERGY E=MC² IN JOULES	POTENTIAL ENERGY MGH S1 TO S2	G=GM₁M₂/R₁₂²KG/M² AT THIS RADIUS
50 to 50 %	5.34×10^{11}	9.0×10^{69} J	8.33×10^{136} J	3.12×10^{72} kg/m²
80 to 20 %	5.34×10^{11}	9.0×10^{69} J	5.69×10^{169} J	5.34×10^{94} kg/m²

TABLE 8.3 Gravity and Potential Energy for 10^{53} kg Split to Produce Two Collapses.

These energies indicate an extraordinary situation with large amounts of energy delivered onto a core of matter.

The energy profile of a black hole as the collapse progresses is depicted in Figure 8.1. In this representation accretion onto the black hole drives the core to promote matter up the particle ladder. This creates a supernova collapse that moves the energy down the well and onto a plateau. The bottom of the well is reached by a further promotion and collapse. The bottom of the well is the maximum stored energy state for the explosive ball of potential energy.

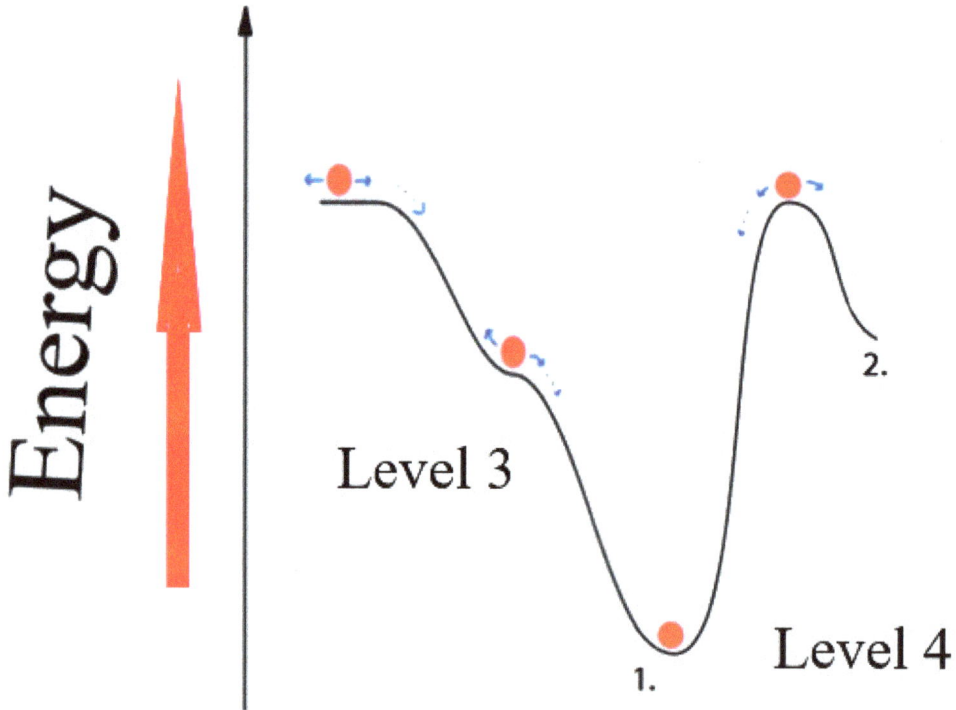

FIGURE 8.1 Energy of Collapse and Explosion.

The compression of the core opens space so that material can rush toward the center. The hot plasma making up the core moves at relativistic speeds due to the exchange of thermal energy. Particles exist in a constant state of wave-function collapse as collisions occur millions of times per second.

Material is produced in this process as gravitational blue shift twists radiation into higher frequency light that forms into matter. This happens as radiation frequencies increase until their mutual attraction takes over and light becomes matter. This is sketched in Figure 8.2 below.

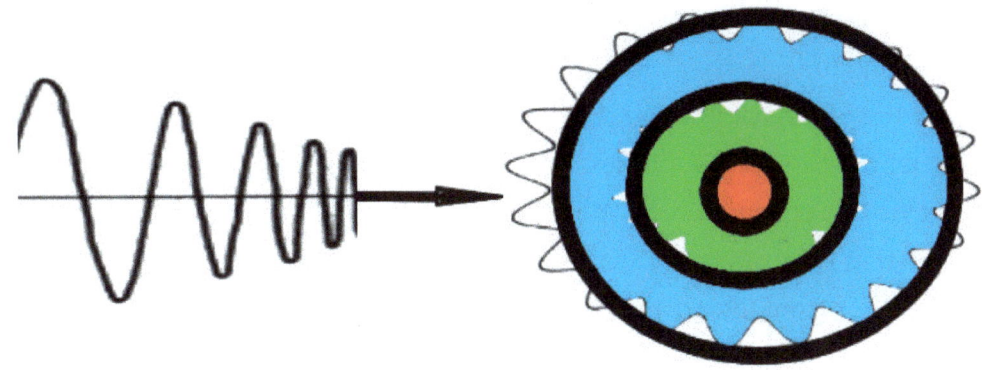

FIGURE 8.2. Light to Matter Transition.

The new matter and any remaining radiation are propelled toward the center of gravity. A growing solid surface of hot quark plasma exists at the center of the black hole. The heat and pressure increase and push matter up the particle ladder. The hot plasma can be a quark anti-quark mix of exotic matter like many more than three particle hadrons. The material at the black hole center therefore acts as an energy manifold storing large amounts of energy in small amounts of space.

BLACK HOLE COLLAPSE

Four types of super nova collapse can be defined for material under gravitational pressure. A Level 1 collapse is one that creates a supernova explosion as described above. The result is a visible body such as a neutron star or a brown or white dwarf star. A Level 2 collapse produces a black hole. The Level 3 collapse happens inside a black hole and is invisible to the exterior universe. A Level 4 Collapse is the one that breaks the gravity well and becomes a universe of stars and galaxies.

A Level 1 collapse can teach us about the other Levels of collapse. This happens when a star depletes its fuel and the supernova explosion leaves a neutron star. The Level 2 collapse happens when the density of material meets the criteria to produce a black hole. The Level 3 collapse allows reorganization of matter within the gravity well. The Level 3 collapse occurs when matter is promoted into higher energy states such as up to charm or top quarks. A black hole can slowly increase in mass and advance toward Level 3 collapse. Matter can step up in energy content as it ascends to a top anti top quark plasma.

The life of a black hole is spent increasing in size and mass as it goes through a sequence of Level 3 collapses. A Level 4 collapse is one that happens only in the cases of very large black holes. This one breaks the gravity well releasing material into the universe. Heating and pressure pushes the scaffold of ascending particles from lowest energy quarks to their highest energy analogues. This creates an increase in the efficiency of the use of space and precipitates a collapse like the one that we record for neutron stars. The up quarks heat into the more energetic charm quarks and a matrix of quark anti-quark mesons can form. This increases the efficient use of three dimensional space. Matter hints about the existence of higher energy particles than the top quark in the Higgs entity. The particles that exist at higher energies than those found at CERN would exist in the hot center of a large black hole. Therefore, we do not know the matter particles above the top quark at 173.1 Gev/C^2 and the Higgs particle at 225 GeV/C^2. It is valid to assume that they exist.

Waves of heat push matter onto a core and promote quarks into higher energy particles. A core that can be multiple light years in diameter acts as a large energy storage manifold. A level 3 collapse is like an avalanche in the mountains that is unheard and unseen. The pressurized core material acts like an ideal gas and matter particles experience millions of collisions per second. When this happens then relativistic mass increase becomes significant. The relativistic mass increase changes the mass and then the Swarszchild Radius. The influence of mass increase is documented in Table 8.4 below.

SPEED FRAC-TION OF C	MASS RELATIVITY 1053 KG AT VELOCITY	GRAVITY AT DISTANCE = 5.34 X 1011 METERS SPHERICAL RADIUS OF 1053 KG TOP QUARK IN N	GRELATIVE/ GNON-RELATIVE	SWARSZCHILD RADIUS R=2GM/C2 2GM/C2	RATIO OF RELATIVE SCR TO NON-RELATIVISTIC
0.000099	Mo(1.00000000490)	1.2491×10^{31}	~1	1.4822×10^{26}	1.01549
0.099	Mo(1.0004855)	1.2496×10^{31}	1.000400	1.5051×10^{26}	1.01549
0.9999	Mo(70.710)	8.8321×10^{32}	70.707	1.063×10^{28}	71.76494
0.999999	M_0 (707.1069)	8.8322×10^{33}	707.06908	1.067×10^{29}	717.677101
0.9999999999	Mo(22,360.6797)	2.7921×10^{35}	22,360.02	3.361×10^{30}	22,695.5673

TABLE 8.4 Mass, Gravity and Swarszchild Radius as Functions of Heating for Massive Black Hole Level 4 Collapse.

The energies for the five different velocities in Table 8.4 are shown in Table 8.5. This is done with top quark masses at 173.1 GeV/C2 where the conversion of kg gives us $3.085787922 \times 10^{-24}$ kg.

SPEED FRACTION OF C	ENERGY $E=MC^2$ JOULES	$KE=1/2$ MV^2 JOULES	ETOT JOULES	MASS IN KG
0.000099	$2.7587181e^{-7}$	$2.7181004e^{-15}$	$2.7587181e^{-7}$	$3.08578e^{-24}$
0.099	$2.7587181e^{-7}$	$2.7194921e^{-9}$	$2.785913e^{-7}$	$3.08728e^{-24}$
0.9999	$1.9506851195 e^{-6}$	$1.9606466e^{-6}$	$3.91133179e^{-6}$	$2.18195 e^{-23}$
0.999999	$1.9506851195 e^{-5}$	$1.9610349e^{-5}$	$3.91172005e^{-5}$	$2.18195 e^{-22}$
0.9999999999	$6.2011428e^{-3}$	$6.2014287e^{-3}$	$1.2402507e^{-2}$	$6.90002e^{-20}$

TABLE 8.5 The Individual Particle Energy as Velocity Increases.

The collapse of material onto the body of the core of a massive black hole brings with it a large amount of energy as evident in Tables 8.3 and 8.5 above. Energy is stored in individual particles as kinetic and rest energies. Here each particle holds 44,957 times the energy in the exited and compressed state than it does in the relaxed non relativistic state. This means the storage capacity for energy in the most heated condition in Table 8.5 is 3.888×10^9 times that of the standard up quark neutronium state. This situation is fluid and can accommodate even more energy by increasing the velocities of the matter particles.

During the lifetime of the black hole a large amount of energy is stored. The excited core receives 10^{170} Joules of energy from the rain of matter (Table 8.3). The energy and gravity tables for the example above suggest the sequence of events to a collapse. The accretion of new matter onto a black hole moves the center of the gravity well to higher temperatures and pressures. The core is compressed and heated causing matter to progress into higher order quarks and store energy in relativistic mass.

The evolution of the black hole progresses over time until it reaches the level 4 collapse. During a collapse the relativistic mass increases pushes the Swarszchild Radius of the black hole outward. Expansion of the Swarszchild Radius brings new material onto the core. This last gasp of the now dying black hole reaches outward into space to extract the last vestiges of its surroundings. The withdrawal of matter toward the core increases the rate of rotation for the body of mass. The faster rotation creates an increase in the centrifugal force pushing outward.

The core of a Level 4 collapse of a super massive black hole is the hottest temperature reached in the history of our universe. The energies in Table 8.5 above show that particle energy for a top quark can go from $2.7587e^{-7}$ to $1.2402e^{-2}$ Joules. The increase in mass at the center is an energy bomb that is ready to explode.

These results indicate that matter can absorb and hold large amounts of energy. Maximum energy is reached when the core reaches its zenith of mass, temperature and pressures. The rain of matter and radiation eventually stops when the expansion depletes all matter and energy from the local volume of space. The end of new kinetic pressure onto the core produces a tipping point in the life of the black hole. Temperature is defined as the capacity for transferring energy from a hot volume to a cold one. The core of the black hole at maximum temperature is much warmer than the surrounding space.

The event that turned a large black hole into our current expansion of matter is known as "The Big Bang". This is proposed to have taken place approximately 13.8

billion years ago. Like a ballerina pulling in her arms the conservation of angular momentum increases the rotation rate.

Turning and turning in the widening gyre
The falcon cannot hear the falconer;
Things fall apart; the center cannot hold;

The Second Coming W.B.Yeats,1919.

This political metaphor works as a description of the tipping point in a massive black hole collapse.

The evolving black hole enters a phase in its natural cycle where competing forces come into play. The collapsing center of mass is increasingly concentrated into a shrinking volume. The spinning rate increases in the conservation of angular momentum while relativistic mass increases. The force of gravity increases as the centrifugal force pushing outward also increases. Two simple equations describe the competing forces. The outward push is the first of the equations below as the centrifugal force. The gravitational attraction is given by the old Newtonian relation with relativistic influence. The velocity when these opposing forces are equal is given by the third equation.

$$F = \frac{M\,v^2}{R} \qquad F = \frac{G\,m_1\,m_2}{r_{12}^2} \qquad v^2 = \frac{G\,M}{R}$$

The changing situation is that the radius decreases while the mass and speed of rotation increase. This increases the centrifugal outward force that opposes the gravitational inward action.

Observation of neutron stars indicates that some rotate at 2/3 the speed of light. The outer mass in a spinning core could approach the speed of light. Taking the 10^{53} kg top anti-top example from Table 8.2 with a radius of 10^{11} M we can estimate the outward force. Here the velocity of equivalence depends upon the mass rotated and would require 10 M/s to expel 100 kg of material. The contest between attraction ad repulsion is superposed upon the supernova collapse that bounces material away from the center. When the massive black hole is broken then the conservation of angular momentum slows the rotation and expels matter into the surrounding space. The evidence that a massive black hole collapse has happened is in the material universe that we observe.

An artist's look at the three stages of a black hole transition into a white hole is in Figure 8.3

FIGURE 8.3. Black Hole Collapse Level 4.

The second panel in Figure 8.3 is representation of the shape of a black hole in collapse. It suggests that the explosive release of material from the core warps the previously smooth surface of the gravity well. This bending of the black hole structure leads to loss of structural integrity and adds to the escape of material. This escape happens at the particle level as matter moving at relativistic speeds flies away from the core of material. Once broken then the black hole losses pressure and further loss of matter follows. The mechanics of the interior of black holes should be a priority in cosmological studies.

The black hole collapse negates any need for faster than light expansion to explain the lack of temperature variance during the early universe. Hot top or higher energy quarks repelling away from a hot core would have a homogeneous energy profile. This would persist as they descended the quark scale falling back to the familiar matter that we know.

BLACK HOLE INTERACTIONS

As stated by Lawrence Krause, "empty space is unstable" and therefore any empty volume of space will evolve to produce a black hole. Quantum fluctuations in the background electromagnetic environment collect due to light upon light attraction. Eventually a massive black hole forms. This is followed by eventual black hole collapse into a matter extrusion like the one that we now occupy.

This means that outside our cosmic microwave background other distant islands of light are currently at some stage of formation. The Copernican Principle reminds us not to conclude that we occupy a special place in the universe. However, in spite of the numerous times that humanity has learned this lesson, the current cosmology holds that we are unique. The current idea is that our 400 billion galaxy universe has exploded into existence as a singular event. The assumption is that there is nothing but empty space on the outside of our Universe. There are also theorists that argue that space and time were created at the big bang. Again this is in direct conflict with the Copernican Principle. The more reasonable answer is that our universe is an island among other islands. This gives answer to the question concerning Dark Energy. Our expanding field of matter and light is pulled by the external gravity field due to other universes like the one we inhabit. The hypothesis therefore that other universes exist outside our current microwave background. Gravity originating from outside our universe influences the expansion of the matter around us. Neutrinos that have travelled from other island universes will pass through the microwave background and be detectable. Also, LEDO type of measurements should be able to eliminate background noise and measure gravitational sources coming from outside. Neutrino and gravitational images of our neighboring matter intrusions would look like our Hubble Deep Space view of distant galaxies. Here each point of gravity would represent a 10^{53} kg black hole equivalent. The state of each source can vary between a concentrated black hole and a field of stars and planets. This is because any gravity source will appear to be a point source from a distance. We can calculate the gravitational attraction between two black hole expansions using the Newtonian approximate and combining powers of ten. In Table 8.6 it is shown that a universe like ours a 100 billion light years or 10^{12} Meters away has a gravitational attraction like our galaxy and the Andromeda galaxy. Again, from table 8.6 when the distance is moved out to 10^{24} meters then the interaction goes down by twenty orders of magnitude. The value goes to a vanishingly small value when the distance is pushed out to a 10^{36} light-years.

MASS/ENERGY OF SOURCES AND TARGET MASS	DISTANCE BETWEEN MASSES IN METERS	GRAVITATIONAL POTENTIAL USING G=6.67 X 10^{-11} $M^3KG^{-1}S^{-2}$ WHERE FORCE IS F=GM_1M_2/R_{12}^2
Earth to an average person M_{earth} (5.97 X 10^{+24}kg) Vs M_{human} (70 kg)	Radius of Earth is 6,371 km	((6.67 X 10^{-11}) times 10^{24} X 10^1)/ (10^7times 10^7)Or 10^{14} / 10^{14} equals 10^1~9.8
Earth to the Moon M_{earth} (5.97 X 10^{+24}kg) to moon (7.35 X 10^{22})kg	Distance of Earth to Moon 384,000 km	((6.67 X 10^{-11}) times 10^{24} X 10^{22})/(10^9times 10^9) 10^{35}/10^{18} Equals 10^{17}
Earth to the sun M_{earth} (5.97 X 10^{+24}kg) mass Sun is (1.99 X 10^{30})kg	Distance of Earth to Sun (149.6 X 10^9) Meters	((6.67 X 10^{-11}) times 10^{24} X 10^{30})/(10^{11}times 10^{11}) 10^{45}/10^{22} equals 10^{23}
Milky Way and Andromeda Galaxies mass of Milky Way and Andromeda at 6 X 10^{11} Solar masses 10^{42}kg	Distance to Andromeda is (2.537X10^6) light years or 10^{22} Meters	(6.67 X 10^{-11} times (10^{42} X 10^{42})/(10^{22}times 10^{22}) or 10^{74}/10^{44} equals 10^{30}
Our Black Hole expansion and another one 100 Billion Light Years Away 10^{53} kg in each of the two matter expansions	Distance between matter expansions is 10^{14} light years or 10^{30} Meters	(6.67 X 10^{-11} times (10^{53} X 10^{53})/(10^{30}times 10^{30})(10^{95}/10^{60}) = Equals 10^{35}meters/second^2
Our Black Hole expansion and another one 10^{24} Light Years Away 10^{53} kg in each of the two matter expansions	Distance between matter expansions is 10^{24} light years or 10^{40} Meters	(6.67 X $10^{-11)}$ times (10^{53} X 10^{53})/(10^{40}times 10^{40})(10^{95}/10^{80}) = Equals 10^{15}meters/second^2
Our Black Hole expansion and another one 10^{29} Light Years Away 10^{53} kg in each of the two matter expansions	Distance between matter expansions is 10^{29} light years or 10^{45} Meters	(6.67 X 10^{-11} times 10^{53} X 10^{53})/ (10^{45}times 10^{45})(10^{95}/10^{90}) = Equals 10^5meters/second^2
Our Black Hole expansion and another one 10^{34} Light Years Away 10^{53} kg in each of the two matter expansions	Distance between matter expansions is 10^{34} light years or 10^{50} Meters	(6.67 X $10^{-11)}$ times 10^{53} X 10^{53})/(10^{50}times 10^{50})(10^{95}/10^{100}) = Equals 10^{-5} meters/second^2
10 m from the center of a black hole that has 10^{53} kg of mass/energy. Assume a small test mass so M2 is negligible contribution.	10 meters	(6.67 X 10^{-11} times 10^{53} X 1)/(10 X 1)(10^{53}/10^1) = Equals 10^{52}meters/second^2

TABLE 8.6 Gravitational Potentials for Different Interactions at Distances (Newtonian Approximation

The interactions that our universe experiences as evident in Table 8.6 tell us that we participate in a web of gravity.

Each expanse of space therefore goes through a cycle as simplified in Figure 8.4.

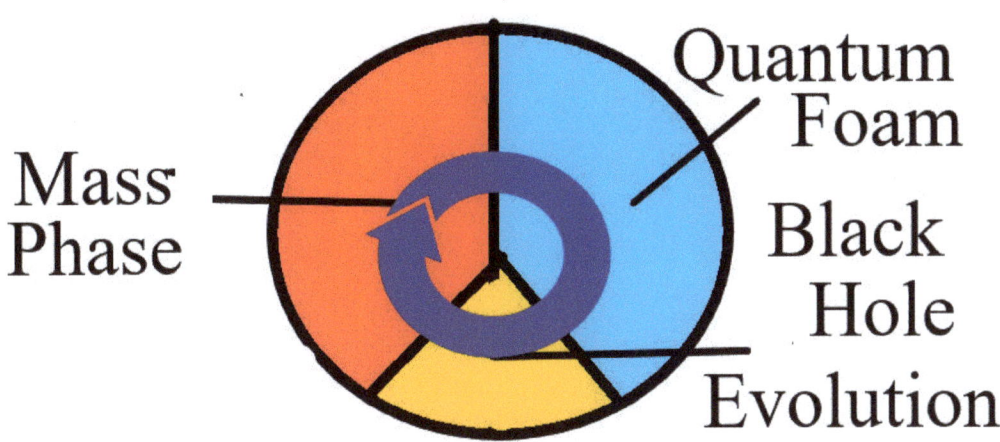

FIGURE 8.4. Space Cycle

The top of the cycle begins with empty space that has low levels of light playing through a three dimensional matrix. Empty space is never absolutely cold at zero degrees Kelvin and is occupied by quantum foam. This evolves until a black hole forms. The black hole evolves to advance to a Level 4 collapse. This is followed by a matter expansion that spreads the energy as matter and light into a vast expanse of space. Matter particles cast adrift in cold space decay into the quantum foam from which they came.

Chapter 9
DARK MATTER, DARK ENERGY, AND THE MEASUREMENT PROBLEM

DARK MATTER

Dark Matter is a form of gravitationally responsive material that does not give off or interact with light. Dark matter was first identified in the 1930's by Fred Swickie. This was achieved by looking at galaxy rotations and concluding the matter count did not correlate with rotation rates. This gravitationally active material was concluded to be a new and exotic form of matter. The current best guess by physicists is that dark matter is a WIMP or "weakly interacting massive particle". This material accounts for five times the mass that is estimated for normal material. The presence of dark matter is crucial in explaining the gravitational interactions that are crucial for galaxy formation. This indicates that dark matter has existed since the 13.8 billion year old formation of our universe.

The current model of matter and our universe suggests that once light becomes matter then it gains mass, charge and spin. This suggests that the gravitational effects that are due to dark matter come from a substance that does not have charge. That substance is however responsive to gravity and so exhibits mass.

There are very few reactions that take place in nature at or near the 100% conversion rate. We can assume that a Level 4 collapse is not fully efficient at converting radiation into matter. All matter interacts with some form of light, but it does not follow that all light interacts with matter. The light that we observe in our universe all comes from matter. However, in the Level 4 collapse of a black hole, light is converted into matter by gravitational blue shift. The conversion process is inefficient and a significant fraction of the black hole mass is blue shifted to frequencies that do not achieve the status of matter. This process produces radiation with frequencies outside the window used by matter. Dark Matter can exist as radiation that is neither absorbed nor emitted by particles. This form of light that has a range of frequencies outside those associated with normal matter. The result is a gravitationally active material that does not form matter and so does not carry charge. The shortest wavelengths achieved by lasers are 10^{-15} M. Without a map of the frequencies in fundamental particle it is impossible to rule out gaps of dark matter. The gravitational evidence of the presence of dark matter is evidence that it exists. However, an experiment that can detect and identify dark matter is needed.

Dark matter makes up some 24% of the total mass energy of our universe. The only approach that allows interaction with dark matter is a light upon light interaction. Here appropriately selected light that can add or subtract by interference with dark matter can be used. The idea is to change the wavelength of the dark matter radiation into frequencies that do interact with matter and so can be measured. Since dark matter is gravitational bound to normal matter, we can use gravity to concentrate this form of radiation. Once isolated and concentrated the dark matter can be irradiated with a range of frequencies. This would be an approach for the modification and identification of this material.

DARK ENERGY

Dark energy is the outward pull upon the matter in our small circle of 400 billion galaxies. Evidence in support of the existence of an outward gravity force is seen in the increase in expansion rate of our universe. The change in expansion rate was discovered in 1998 by the Hubble telescope. A possible explanation for this is that surrounding islands of matter, like our own, produce a gravity field. This disposes of any need for the proposed anti gravity field and normal gravity explains all. The experimental result that our universe expands at an increasing rate is consistent with there being a mass pulling our universe apart. Our local expansion rate increases as internal attraction is weakened while the external pull increases or remains constant. Distances inside our 9.6×10^{10} light year diameter expanse are small relative to the more than 10^{20} light year distance to other islands. This fact of nature is evidence for the presence of matter outside our CMB.

Alternate universes would have the same gravitational affect upon us whether they are spread in a wide diameter or concentrated into a single black hole. The same laws of nature would exist in these alternate swirls of energy since they would be like our small circle of matter. The same light based quarks, leptons and atoms would exist there as we have here. These universes would experience the same cycle as our own universe. This means they would experience the same cycle as in Figure 8.4 above. Each cycle hits the two extremes of order and disorder as entropy toggles between two extremes.

The hypothesis by Hubble that galaxies are receding away from one another has lead cosmologists to explain this as expanding space. Here, it is argued that since all galaxies are receding away from one another, empty space between galaxies must be growing. This conclusion fails to provide any mechanism to explain how empty space can increase its volume. This is reminiscent of the Minkowski oxymoron that empty space is mutable. The material or light that occupies space can be manipulated. The interpretation that space is mutable can be mathematically equivalent to the warping of empty space. However, a better explanation is available.

The gravitational pull from outside our universe pulls on light to create a red shift. This is true for all light that moves inside our collection of galaxies. The observation is that the material in the universe is expanding away from us at all points in space. This is explained if an external mass red shifts all light coming to us. This does not mean the universe is not expanding, but that we do not know the expansion rate. This means that we lack a correct answer to the question of the size and age of our universe. The outward pull should increase all red shifts making our expansion appear to be faster than the correct value. The addition of a shift from a genuine expansion and a red shift due to external gravity gives the correct red shift. A result of the addition of two contributions explains how galaxies can appear to recede at greater than light speed.

Therefore, empty space does not grow between galaxies. Exposure to the gravitational attraction from outside produces a red shift that correlates with the distance travelled and this is observed. The addition of a gravity shift and velocity shift explains why a red shift for some galaxies can exceed the speed of light. The result is that our current determinations of the age of our matter intrusion are wrong.

Calculations of the gravitational potential in Table 8.6 suggest that distance of 10^{29} to 10^{33} Light Years could be close enough to red shift our light. The increasing expansion rate for our universe is explained by a weakening of internal gravitational attraction. The external pull remains constant while the internal attraction falls off with distance.

MEASUREMENT PROBLEM

The measurement problem in modern cosmology is in the divergent results when different methods are used to measure distances. Geometric parallax is the method of measuring the difference in direction from two separated positions. An astronomical object observed from two different directions can yield its distance if the intensity is known. The positions of the earth at opposite sides of its orbit are used to measures distances. Parallax requires a known intensity and cepheid variables have a periodic variability correlated with their intensities. This gives a known intensity that permits determination of distance from earth. The second method of determining the distance to objects in space is by determining the Hubble Constant or red shift of light coming from any object. The premise here is that empty space is expanding and thus creating a red shift from objects receding from earth. The measurement problem is that these methods give different results.

An explanation for the measurement problem comes from the above description of dark energy. The gravitational pull from outside our universe has an influence upon any light travelling inside. The red shift created by this gravitational influence changes the measured red shift described by the Hubble Constant. This can be used as a tool used to determine the strength of the gravitational attraction pulling our universe apart. The distance to other universes can be measured by getting the change in our expansion rate over time.

Chapter 10
UNIVERSE

Our journey has brought us to a place where the model of matter can take us no further. This is the part of our universe where experiment cannot take us due to distance and time constraints. The objective is to probe the universe that exists outside our current microwave background. The journey into a wider realm of space requires a new vehicle. Our next transport requires the invention of a vehicle that can exit our universe and grow in size. The H.G.Wells tale of The Time Machine has been mentioned above with respect to space-time. The vehicle we require is inspired by the time machine but this transport moves through the size dimension. Therefore, like the time machine, this mythical vehicle has dials that can set a size dimension to visit.

The idea is that both matter and antimatter universes may exist outside our universe. These may be gravitational connected to one another and this presents an alternate form of interaction between islands of matter and light. The Copernican Principle tells us that we are not unique in the universe. The Copernican lesson may also apply to size dimensions. The range of size that we know stretches from the Planck scale at 10^{-34} M to the current 98 billion light year diameter of our universe. The diameter is therefore 10^{26} M and the size range represents 60 orders of magnitude. In a universe of very large size it may be possible that a range of 60 orders of magnitude exists elsewhere. This means that between 100 and 200 orders of magnitude above our domain there may be organization like the one we know.

Dimensions few hundred orders of magnitude above our own are the first stop in the size machine. The view of our universe from outside the CMB is similar to the one from inside as the CMB is opaque. Moving in size away from our universe we look at an opaque pearl of light. Looking around the neighborhood through the gravity lens it is possible to see other balls of light spread through space. In our view matter expansions spread away from us like stars spread through a dark night.

Moving further away from our star field of universes, other matter expansions appear as filaments and strands. There can be a size dimension where energy collects into clusters of matter and light that resemble the particles that we know. Are we a particle making up a grain of sand on a distant beach? We must accept that a different range of 60 orders of magnitude could entertain organized light, matter, chemistry and life.

A thousand or a million orders of magnitude away from home would produce the possibility of many levels of organization. This idea is reminiscent of the ancient fable in which it was asserted that the earth rested on the back of a giant turtle. The natural question that followed was, well what is that turtle standing upon? The easy answer came that this turtle was standing upon the back of a turtle and it was turtles all the way down. This old fable may have a metaphoric wisdom not immediately obvious. These alternate dimensions of size are unavailable if existent.

All objects can only exist in three dimensions if they are made from light. This follows because light extends itself in three dimensions as in Maxwell's Equations. One can convince oneself that it is not possible to make four dimensional objects with three dimensional building blocks. This is done by dropping one axis and using two dimensional blocks to make a three dimensional figure. This is modeled in Figure 10.1 below.

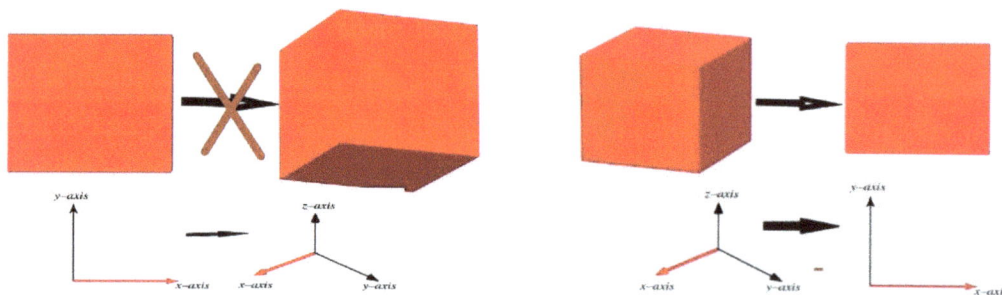

FIGURE 10.1 Square to Cube is Impossible, Cube to Square Allowed

Consider a block that has extension in the XY plane and zero value along the Z axis and try to make a cube with extension in all three axes. Conversely we can set one of the axes infinitely close to zero and represent two dimensions with a three dimensional object. This means that any object having more than three dimensions cannot be made from light. The implication is that all other universes will be three dimensional like the one we have.

Time in any dimension is contingent upon two factors including distance and the speed of light. It therefore follows that other dimensions would have rates of time passage based upon the range of their sizes.

Taking a perspective from a time and dimension a million orders of magnitude away from our own is sobering. Time merges the millions killed by history's wars with those about to die. We do not need another 'great step forward'. We can do without another 'final solution' or any 'military operation'. Finding ourselves in a natural universe huge with opportunity and challenge is breathtaking. This knowledge hints at the journey our progeny will discover. Looking back from a universe that is magnitudes away from ours brings a new perspective. A traveler in a distance and foreign land dreams in *Pictures of Home*.

EPILOGUE

When home is achieved then a journey will have changed the place long since left. In this case our universe has a new shine. This voyage of discovery has revealed the properties of matter that make our universe work. Mystical Quantum Mechanics is exposed. The cycle that causes empty space to become a black hole and then a universe of matter is revealed. Matter expansions outside our cluster of galaxies exist and occupy a larger universe. We have realized our inheritance of a large and comprehensible universe.

References

On the Origin of Species by Means of Natural Selection, by Charles Darwin, M.A. London: John Murray, Albemarble Street, 1859,

Relativity: The Special and General Theory Albert Einstein, © 1920 Publisher: Methuen & Co Ltd First Published: December, 1916

The **Feynman Lectures on Physics** is a physics textbook based on some lectures by Richard Feynman, The lectures were presented before undergraduate students at the California Institute of Technology (Caltech), during 1961–1963.

The Strange Theory of Light and Matter by Richard P. Feynman, Princeton University Press 1986.

A Teatis on Electricity and Magneticism; James Clerk Maxwell, 1873 Clarington Press,, Oxford.

The Time Machine, by H. G. Wells 1895,London, Penguin Classics, 2012

The Trouble With Physics: The Rise of String Theory, The Fall of a Science, and What Comes Next Paperback – Illustrated, Sept. 4 2007 by Lee Smolin.

A Universe from Nothing: Why There is Something Rather Than Nothing written by Lawrence M. Krauss. 2012, Publisher, Free Press imprint of Simon & Schuster,

Electromagnetic Mass, Relativity, and the Kaufmann Experiments, American Journal of Physics 49, 1133 (1981); https://doi.org/10.1119/1.12561

Lorentz, H.A. (1900), **Über die scheinbare Masse der Ionen (On the Apparent Mass of the Ions)** Physikalische Zeitschrift, 2 (5): 78–80, Lorentz, Hedrik Antoon (1904), **Electromagnetic Phenomena in a System Moving With Any Velocity Smaller Than That of Light,** Proceedings of the Royal Netherlands Academy of Arts and Sciences, 6: 809–831, Bibcode:1903KNAB....6..809L

Attempt of a Theory of Electrical and Optical Phenomena in Moving Bodies , Lorentz, Hedrik Antoon (1895), Leiden: E.J. Brill, Bibcode:

Relativity: The Special and General Theory, New York (1916) :by Albert Einstein, H. Holt and Company.

Treatise on Electricity and Magnetism, by James Clerk Maxwell MA, 1873, Published by Macmillan and Company".

Properties of High-Density Matter in Neutron Stars by by F. Weber et al, G. A. Contrera, M. G. Orsaria, W. Spinella, O. Zubairi

August 2013 Modern Physics Letters A 29(23)